WA 1300908 7

D1381046

ONE WEEK LOAN
UNIVERSITY OF GLAMORGAN
LEARNING RESOURCES CENTRE

Renew Items on PHONE-it: 01443 654456
Help Desk: 01443 482625 Media Services Reception: 01443 482610
Items are to be returned on or before the last date below

Digital Terrain Modeling

Acquisition, Manipulation, and Applications

For a listing of recent titles in the *Artech House Remote Sensing Library*, turn to the back of this book.

Digital Terrain Modeling

Acquisition, Manipulation, and Applications

Naser El-Sheimy
Caterina Valeo
Ayman Habib

ARTECH
HOUSE

BOSTON | LONDON
artechhouse.com

Library of Congress Cataloging-in-Publication Data
El-Sheimy, Naser.
    Digital terrain modeling: acquisition, manipulation, and applications/Naser El-Sheimy, Caterina Valeo, Ayman Habib.
    p. cm.—(Artech House remote sensing library)
    Includes bibliographical references and index.
    ISBN 1-58053-921-1 (alk. paper)
    1. Digital mapping—Methodology.   I. Valeo, C. (Caterina) II. Habib, A. (Ayman) III. Title. IV. Series.

    GA139.E49 2005
    526—dc22   2005045272

British Library Cataloguing in Publication Data
El-Sheimy, Naser
    Digital terrain modeling: acquisition, manipulation, and applications.
    —(Artech House remote sensing library)
    1. Digital mapping   2. Remote sensing   3. Relief models—Computer simulation
    I. Title II. Valeo, C. (Caterina)   III. Habib, A. (Ayman)
    526

ISBN-10: 1-58053-921-1

Cover design by Igor Valdman

International Standard Book Number: 1-58053-921-1

10 9 8 7 6 5 4 3 2 1

Contents

Preface

This book is the product of several years of effort devoted to educating geomatics engineering students in digital terrain modeling at the University of Calgary. Dr. Naser El-Sheimy began developing course notes that gradually evolved into this book with the assistance of Drs. Valeo and Habib. However, pedagogy was not the only motivation behind the development of this book. The authors also wanted to fill a very large gap in the literature on digital terrain models (DTMs) by writing a book that covers all the essential elements of DTMs and the applications in environmental modeling and mapping in a mathematically rigorous, highly advanced, thoroughly complete, and cohesive manner.

This book is intended for a wide variety of users both within and external to the geomatics community. It has been written to not only provide a rigorous treatment of DTMs for those interested in advancing the field, but also to provide readers with the possibilities in DTM applications, thus making it accessible to many. Some background in geomatics engineering and mathematics are all that is required to fully understand all facets of this work. This book consists of eight separate chapters and together they provide the reader with an understanding of the essential elements of DTMs and their meaning.

Chapter 1 provides a general overview of the utilities of DTMs with a brief introduction to the terminology and the basics of the data models used. At the heart of this chapter is a description of the five elements of DTMs: generation, manipulation, interpretation, visualization, and application. The user will come to realize why DTMs are so important to so many fields, including civil engineering, earth sciences, planning and resource management, remote sensing, mapping, and military applications.

The second chapter discusses the first element in greater detail: the process of generating a DTM. An extremely useful chapter, the material covers the primary mechanisms of DTM generation including raster to vector conversion from cartographic data sources and the processes involved, generating grids from contour data, photogrammetry, LIDAR, terrestrial laser scanners, and interferometric SAR. The user will learn the essential components of each of these approaches as well as the benefits and disadvantages. In addition, a unique bonus of this chapter is the detailed exploration of the technology of "direct georeferencing." This innovative and cost-effective method is revolutionizing the essential task of georeferencing in new DTM data acquisition techniques. Chapter 3 further discusses the element of DTM generation by detailing information on data structures. Here the finer points of working with GRIDs, TINs, contours, lattices, or irregularly spaced points are provided.

Chapter 4 presents the next step in DTM analysis: manipulation. Data manipulation is devoted to describing the preanalysis phase that often refines the DTM to suit application needs. This may include editing, filtering, enhancing, merging, joining, and elevation resampling. The reader will learn the various interpolation techniques that exist and their advantages and disadvantages. This chapter naturally leads into Chapter 5, which provides a closer look at one of the preferred methods of interpolation: KRIGING. Complete, in-depth mathematical developments of the various forms of KRIGING are provided and the reader will come to understand how the various methods work and when and why to use each.

Chapter 6 provides vital information on the subjects of DTM generalization and quality control. Both are important components of DTM manipulation and interpretation. The reader will become familiar with alternatives for generalizing DTMs together with the impact of different generalization methodologies. The chapter also discusses various factors affecting the quality of the DTM as they pertain to different data acquisition systems. Finally, several methodologies for quantitative assessment of the quality of the generated DTM will be presented.

Chapters 7 and 8 describe the visualization and application elements of DTMs. These chapters cover a wide range of techniques for "visualizing" and "exploring" DTMs to provide additional information leading to improved decision-making. Chapter 7 begins with visualization techniques ranging from simple shading to data fusion methods that involve draping orthophotos and the mathematics involved. Orthophotos are discussed in detail along with the rectification process. The reader is also exposed to common exploration techniques such as first- and second-order derivatives commonly used in environmental decision-making.

Chapter 8 describes applications that are at a level above those given in Chapter 7. Areas of environmental and water resources engineering use DTMs extensively and the nature of this use is described in this chapter. The reader will become familiar with the implications of DTM use, the limitations, and the possibilities in areas such as drainage modeling, erosion modeling, risk modeling during fires and avalanches, and the implications of using grids or TINS. Issues such as error and resolution and their impacts on derived model parameters are discussed.

The authors recognize that this book could not have been produced without the assistance of several key people. We would like the thank Dr. John Bossler and Dr. Husam Kinawi for their technical reviews and valuable feedback. We would also like to thank the anonymous reviewer provided by Artech House for his/her comments and suggestions. In addition, the authors would like to thank Ms. Monica Barbaro, Mr. Cameron Ellum, Dr. Sameh Nassar, Mr. Mwafag Ghanma, Mr. Chang-Jae Kim, and Ms. Rita Cheng for their help in editing this book. We would also like to thank our families for their continued support of our efforts. Thanks are also due to the Department of Geomatics Engineering for its support of the development of this book.

Finally, the authors would like to express their gratitude to a number of companies and agencies who provided figures shown throughout this text. They are listed as follows:

AltaLis
Applanix Corporation (now a subsidiary of Trimble)
BAE Systems
iMAR GmbH
Intergraph Corporation
Intermap Technologies Corporation
Natural Resources Canada
Optech Incorporated
PCI Geomatics
RIEGL Laser Measurement Systems GmbH
TopoSys, Topographische Systemdaten GmbH
3D Laser Mapping Ltd

Chapter 1

Introduction

1.1 WHAT IS DIGITAL TERRAIN MODELING?

The concept of creating digital models of the terrain is relatively recent and the term digital terrain model (DTM) is generally attributed to two American engineers at the Massachusetts Institute of Technology during the late 1950s. The definition they coined then was: a "DTM is simply a statistical representation of the continuous surface of the ground by a large number of selected points with known X, Y, Z coordinates in an arbitrary coordinate field" (Miller and LaFlamme, 1958). Specifically, their early work was concerned with the use of cross-sectional data to define the terrain. Since then, several other terms, such as digital elevation model (DEM), digital height model (DHM), digital ground model (DGM), and digital terrain elevation data (DTED) have been combined to describe this and other closely related processes. Although in practice these terms are often presumed to be synonymous, in reality they often refer to distinct products (Petrie and Kennie, 1990). The following discussion is therefore provided in an attempt to simplify and standardize the use of these terms.

By reviewing various dictionaries and technical literature, a variety of definitions or meanings for the words and terms used in DTM are encountered. For example, ground may be defined as "the solid surface of the Earth," "a solid base or foundation," "a surface of the Earth," or "a portion of the Earth's surface." Similarly, height can have different meanings such as "measurement from base to top," "distance upward," or "the elevation above the ground or a recognized level." Elevation can be "the height above the horizon" or "the height above a given level, especially that of the sea." Finally, terrain has the rather different concept or meaning of being "a tract of country considered with regards to its natural features," or "an extent of ground, region, territory, etc." (Petrie and Kennie, 1990).

From these definitions, some differences begin to manifest themselves between different terms:

1

1. DEM: The word elevation emphasizes the measurement of height above a datum and the absolute altitude or elevation of the points in the model. DEM as a term is in widespread use in the United States, and generally refers to the creation of a regular array of elevations, normally squares or a hexagon pattern over the terrain.
2. DHM: This is a less commonly used term with the same meaning as DEM since the word elevation and height are normally regarded as synonymous. The term seems to have originated in Germany.
3. DGM: This term seems to lay its emphasis on a digital model of the solid surface of the Earth. In this way, there is presumed to be some connection between elements that are no longer considered discrete. This connection generally takes the form of an inherent interpolation function that may be used to generate any point on the ground surface. The term is in general use in the United Kingdom, although its use has to some extent been superseded by DTM.
4. DTM: A more complex concept involving not only height and elevations but also other GIS features such as rivers and ridge lines. Moreover, DTM may also include derived data about the terrain such as slope, aspect, and visibility. In a narrow sense, a DTM represents terrain relief. In its general form, a DTM is considered by most people to include both planimetric and terrain relief data.
5. DTED: A term used by the U.S. Defense Mapping Agency (DMA). It describes essentially data produced by the same process although it specifically uses grid-based data.

The two most widely used terms are DEM and DTM. These terms are often used synonymously. Usually, the acronym DTM refers to the altitude of the ground itself. On the contrary, a DEM includes the maximum altitude everywhere (including roofs of buildings and tops of trees). Therefore, models generated from digitizing topographic mapping (see Chapter 2) will produce a DTM, while those derived from satellite imagery will be DEMs. On the other hand, elevation data derived from new sensors such as LIDAR (LIght Detection And Ranging; see Chapter 2 for further details) can be processed to produce either DEMs or DTMs. In this book, the term DTMs will be used as a general term for representing the different acronyms discussed above.

1.2 THE IMPORTANCE AND NEED FOR DIGITAL TERRAIN MODELS

The scientific community and the commercial market are increasingly aware of the importance of DEMs in their applications. There are a large number of military, environmental, engineering, and commercial GIS applications that rely entirely on the ready availability of digital terrain databases. Their earliest use dates back to the 1950s (Miller and LaFlamme, 1958) and since that time they have proven to be an important method for modeling and analysis of spatial-topographic information. The application domains of DTMs include the following:

1. Civil Engineering: The range of civil engineering projects where DTMs are used is almost endless, covering roads, railways, dams, reservoirs, canals, landscaping, land reclamation, and mining. The range of facilities included in the terrain modeling permits the survey and design of projects mentioned above. Civil engineers are mainly interested in using DTMs for cut-and-fill problems involved with road design, in-site planning, setting-out information, 3-D landscape modeling, visualization for civil engineering tasks, location and route planning, and volumetric calculations in building dams, reservoirs, and the like. It may be pertinent to point out that owing to such overt concerns with volume and design, calling a DTM a "terrain model" has more relevance to a civil engineer than other DTM users.

2. Earth Sciences: Exact information about the Earth's surface is of fundamental importance in all geosciences applications. For example, in weather forecast and climate modeling, models of conversion processes between the ground and the atmosphere as well as of movements in the lower atmospheric strata also rely on uniform and global DTMs. Many other Earth or geoscientific applications center mainly on specific functions for modeling, analysis, and interpretation of the unique terrain morphology. These may include climate impact studies, geological and hydrological modeling, geomorphology and landscape analysis, biophysical modeling analysis, geological studies, generating hazard maps (seismic hazards, landslide hazards, volcanic hazards, beach erosion/accumulation hazard), drainage basin network development and delineation (see Figure 1.1), hydrological runoff modeling, geomorphological simulation and classification, and geological mapping. Generating slope and aspect maps, and slope profiles for creating shaded relief maps are popular tasks performed in the earth sciences that employ DTMs.

3. Planning and Resource Management: This is a major grouping of diverse fields including remote sensing, agriculture, soil science, meteorology, climatology, environmental and urban planning, and forestry, whose central focus is the management of natural resources. Applications best characterizing this domain include site location, support of image classification in remote sensing by DTM derivatives, the geometric and radiometric correction of remote sensing images, soil erosion potential models, crop suitability studies, and wind flow and pollution dispersion models. As is evident, this group covers a wide range of concerns, and concomitantly requires a matching range of functionality. These include procedures for data capture, editing and verification, established data models and structures in both raster and triangulated irregular networks (TIN) domains, and robust analytical, modeling, and visualization tools.

4. Remote Sensing and Mapping: In remote sensing and mapping, DTMs are used, together with GIS, to correct images or retrieve thematic information with respect to sensor geometry and local relief to produce georeferenced products. Thus, for the synergic use of different sensor systems (and GIS), DTMs are a prerequisite for georeferencing satellite images and correcting terrain effects in photogrammetric and radar scenes. On the other hand, representations of digital elevation data can be greatly improved visually by draping satellite imagery over the terrain creating realistic 3-D views (see Figure 1.2).

5. Military Applications: The military is not only a leading consumer of DTMs, they are also a significant producer. Almost every aspect of a military environment depends on a reliable and accurate understanding of the terrain, elevation, and slope of the land surface. The military use of DTMs combines facets and methods of all the previous application domains, and their end objectives are very specialized and demanding. Examples of such use would include intervisibility analysis for battlefield management, 3-D display for weapons guidance systems and flight simulation, and radar line-of-sight analyses.

Figure 1.1   A 3-D view of a derived drainage network (derived from Autodesk Envision).

1.3 ELEMENTS OF DIGITAL TERRAIN MODELING

The production and use of DTMs generally involves five discrete procedures or tasks. These are generation, manipulation, interpretation, visualization, and application, as shown in Figure 1.3. Deriving products from a DTM should not be viewed as a one-way process, but rather as the result of various interrelated stages

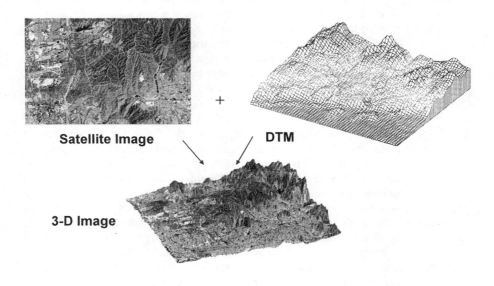

Figure 1.2 Example of grid rendering using satellite imagery.

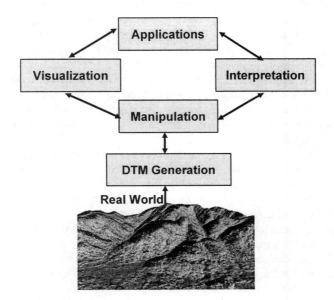

Figure 1.3 Main tasks of DTM.

in modeling. For example, a DTM may be modified by model manipulation procedures. It might then be displayed by visualization procedures, or analyzed through interpretation functions. Visualization and interpretation in turn may require or support further modification or adaptation of the original DTM. Thus, results of individual modeling steps may feed back into previously run procedures. The following section will describe each of these tasks.

1.3.1 DTM Generation

DTM generation (i.e., building the DTM) forms the basis for all subsequent operations. It consists of two subtasks, namely the measurement and digitization of original terrain observations (terrain data capture) and the formation of relations among the diverse observations to build DTMs (model construction). There are a number of choices when it comes to the generation of DTMs, and the preferred option is always going to be a balance between the desired accuracy of the DEM and the costs involved in its creation. Elevation data ranges from free, low resolution, low accuracy products (e.g., GTOPO30, - a global digital elevation model with a horizontal grid spacing of 30 arc seconds) through more costly medium resolution products (derived from satellite data), to high-accuracy high-resolution models typically derived from airborne sources (LIDAR, photogrammetry).

DTMs derived from contours are perhaps the most common. This is because digital contour data has been developed from analog maps for the longest time when compared with the other sources. Contours from analog maps can be digitized by manual digitizing, semi automated line-following, or automatic raster-scanning. The contours created from digitizing are then registered, vectored, edited, and tagged with elevation values.

Photogrammetric data capture encompasses two methods for deriving remotely sensed data: aerial photography and digital satellite imagery. Both of these methods have been used to generate accurate high-resolution topographic data. There are also a number of newer sources, such as radar and laser altimetry and synthetic aperture radar (SAR). Interferometry is also being developed for deriving topographic data.

Data derived from surveying is the most accurate available, as the operation is designed to draw the most information from the area. This is done by incorporating a significant number of points on the terrain. Such high level of accuracy, however, requires a lot of time in the field, and therefore ground surveys are only feasible for specific projects involving small areas.

The fundamental architecture of a DTM derives from the data model used to represent it. DTMs can be represented either by an image method (using a point or line model), or mathematically. A variety of data structures have been tried and tested for storing and displaying topographic surfaces; however, two in particular have been the most popular and best explored: the rectangular grid or elevation matrix structure, and the triangulated irregular network (TIN) structure. Both are image representations that use a point model. The elevation matrix or rectangular grid is the most commonly used modeling construct for a DTM. This is because the data structure of a grid shares much similarity with the file structure of digital computers—elevations can easily be stored in a computer as a two-dimensional array (every point can be assigned to a row and a column). This also streamlines both information processing and algorithm development. The TIN model provides a network of connected triangles with irregularly spaced nodes or observation points with x, y coordinates and z values. Its major advantage over an elevation matrix is its ability to generate more information in areas of complex relief, and avoid the problem of gathering redundant data in areas of simple relief.

1.3.2 DTM Manipulation

Manipulation procedures are essential in order to modify, calibrate, or refine existing DTMs. DTM manipulation involves the modification and refinement of DTMs and the derivation of intermediate models. It include editing, filtering, merging, and joining DTMs, as well as procedures for converting DTMs from one data structure to another (e.g., TIN to grid). Editing is required to correct errors and update DTMs. Filtering involves smoothing and enhancement functions by using low-pass and high-pass filters, respectively. Smoothing reduces or removes details in the DTM while enhancement highlights terrain discontinuities. Filters can also be used to reduce the data volume of a DTM, which in turn can save disk storage space or economize processing time. Filtering can also be done to aggregate or resample the image resolution. DTMs, when they are adjacent, may be merged together while joining gridded DTMs is possible only when orientation and resolution are compatible.

1.3.3 DTM Interpretation

The value of a DTM is a function of its extractive knowledge, its attributes, and information regarding terrain. DTM interpretation encompasses the analysis of DTMs in order to extract information from DTMs, for use in GIS application models or to support other terrain modeling tasks. Interpretation serves to provide such geomorphological information through quantitative analysis of the digital terrain. Such analyses can provide direct input to a range of environmental resource management applications. Interpretation may be directed at two different levels: general geomorphometry and specific geomorphometry. General geomorphometry aims to generate slope and its dual derivatives of gradient and aspect while specific geomorphometry analytically portrays terrain features and attributes in relation to the surface hydrology. This is extremely useful in drainage basin delineation and hydrological runoff simulation, as well as geomorphological modeling studies.

1.3.4 DTM Visualization

Visualization of DTMs plays an essential role in their perceptual understanding and appreciation. It focuses on the display of DTMs as well as on information derived from DTMs. Visualization serves the purpose of communicating modeling results to the specialist as well as the end user, and may directly support decision making through visual interpretation. Visualization can take two forms:

1. Interactive visualization for exploring the model, calibrating, and refining its premise;
2. Static visualization, for basic communication of the results.

The best-known methods for displaying and interpreting DTM data, like contours and shaded relief, have aided visualization for a long time notwithstanding their broad degrees of abstraction. There has been a lot of progress in hardware and software related to visualization in areas like 3-D modeling, photo-realistic scene rendering, and animation. All of this is beneficial to interpretation operations since visual analysis of graphical representations is closely linked to it. Figure 1.4 shows, for example, a vector projected on a DTM.

Figure 1.4   Projecting vectors onto the DTM.

1.3.5 DTM Application

DTM application forms the context for digital terrain modeling. DTM application involves the development of appropriate application models in relevant disciplines in order to make efficient and effective use of the available technology. Of course, each particular application has its specific functional requirements relative to terrain modeling techniques. Based on these requirements, the quality of existing methods can be assessed.

DTMs have developed to a point where they provide the core functionality for a variety of applications. In fact, it could be argued that all applications including geographic data require elevation data. Technological advances in computer graphics and display, spatial theory, databases, and a host of related areas are making it possible to explore and apply DTMs in varied applications. Broadly, these include the applications discussed in Section 1.2.

1.4 DATA MODELS IN DTM

During the process of acquiring terrain data, a relatively unordered set of data elements is captured. In order to construct a comprehensive and usable DTM, it is necessary to establish the topological relations between the data elements as well as an interpolation model to approximate the surface behavior.

A continuous surface, such as the Earth's, has an infinite number of points that can be measured. Obviously, it is impossible to record every point, and consequently a sampling method must be used to extract representative points. These discrete points can then be used to build a surface model that approximates the actual surface. The type of surface model dictates the sampling method. A surface model should (ESRI, 1992):

1.   Accurately represent the surface;
2.   Be suitable for efficient data collection;
3.   Minimize data storage requirements;
4.   Maximize data handling efficiency;
5.   Be suitable for surface analysis.

Three methods are commonly used to represent surfaces in digital form: contour lines, grids (lattice or elevation matrix), or triangulated irregular network (TIN).

1.4.1 Contours

Contour or isolines of constant elevation at a specified interval are probably the most familiar representation of terrain surfaces. Contour maps for most of the world are readily available at a variety of scales. Contour accuracy depends upon whether the isolines have been generated from primary or derived data. When contours have been captured directly from aerial photographs as primary data using a stereoplotter, the contours are highly accurate. If the contours have been generated from point data, the location of the contours must be interpolated between known values. One of the major drawbacks of contours is that they only indicate surface values along the isolines. Surface anomalies between contour intervals cannot be represented. Once the surface has been represented as contours, interpolation can be used to derive an elevation for locations between contours.

When presented on hardcopy maps, each contour is drawn as a continuous line following the contour interval across the surface. Each line contains an infinite number of potential sample points. When digitizing a map to automate its contours from hardcopy to digital format, the lines must be sampled again because it is impractical to store every point along a line. The most common technique is to digitize or select vertices at significant inflection points along the contour, such that the digitized line does not significantly deviate from the actual contour. The process of automating contours from hardcopy manuscripts is explored in more detail in a later chapter.

1.4.2 Grids

Grids are a matrix structure that implicitly record topological relations between data points. Since this data structure is similar to the array storage structure of digital computers, the handling of elevation matrices is simple, and hence, grid-based terrain modeling algorithms tend to be relatively straightforward. On the other hand, the point density of regular grids cannot be adapted to the complexity of the relief, so that an excessive number of data points are required to represent the terrain to the required level of accuracy. There has been, however, extensive research in the use of the quad tree (see Chapter 3) methods to handle digital elevation data in grid form so that data redundancies can be reduced.

A lattice is the surface interpretation of a grid represented by equally spaced sample points referenced to a common origin and a constant sampling distance in the x and y directions. Each mesh point contains the z value of that location, and is referenced to a base z value, such as mean sea level. Surface z values of locations between lattice mesh points can be approximated by interpolation between adjacent mesh points.

In a lattice, each mesh point represents a value on the surface only at the center of the grid cell; it does not imply an area of constant value. In contrast, categorical grids consider each grid cell as a square cell with a constant attribute value. All locations within the grid cell are assumed to have the same z value. However, categorical grids or their associated attributes can be interpreted as a lattice surface. Figure 1.5 shows the representation of elevation in lattice and categorical grids.

The accuracy of an array in representing the surface is dependent upon the distance between sample points. Because the sample points are at a fixed interval, surface features such as streams and ridges cannot be represented directly by the sample array. Similarly, surface anomalies such as peaks and pits can be missed. The only way to increase the representation of these features is to increase the lattice resolution (i.e., reduce the distance between sample points).

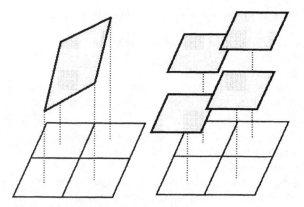

Figure 1.5 Lattice versus categorical interpretation.

The accuracy of an array as a surface representation can be increased by decreasing the interval between sample points, noting the following:

1. A large sampling interval may miss important variations in the surface.
2. Increasing the lattice resolution by decreasing the sampling interval usually results in a more accurate surface representation.
3. Decreasing the distance between sample points can be at the expense of increased data redundancy. This is particularly true for areas of the surface that do not exhibit significant variation, yet have many more mesh points than are needed to accurately represent them.

The effect of oversampling and undersampling is shown below in Figure 1.6.

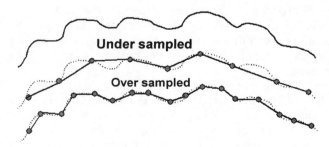

Figure 1.6 Effect of grid size on surface representation.

Grid Data Structure

Advantages:
  1. Already have grid DEM with no further processing;
  2. Suitable for trend surfaces;
  3. Easy to store and manipulate;
  4. Easy to integrate with raster databases;
  5. Has more natural appearance of derived terrain features.

Disadvantages:
  1. Inefficient sampling, although progressive sampling on increasingly finer grids according to relief complexity is possible (however, there will always be redundant points);
  2. The highest or lowest points on the landscape are rarely sampled, since they are not likely to fall directly on the sample grid;
  3. Inability to use various grid sizes to reflect areas of different complexity of relief.

1.4.3 Triangulated Irregular Network (TIN)

The triangulated irregular network (TIN) model is a significant alternative to the regular raster of a DTM and has been adopted by numerous GIS and automated mapping and computer packages. The TIN model was developed in the early 1970s (Peucker et al., 1978) as a simple way to build a surface from a set of irregularly distributed points. Several prototypes using this data structure were developed in the 1970s and commercial systems using TINs began to appear in the 1980s as contouring packages (sometimes embedded in GIS).

Irregularly distributed sample points can be adapted to the terrain, with more points in areas of rough terrain and fewer in areas of smooth terrain. An irregularly spaced sample is therefore more efficient at representing a surface than a regularly spaced sample such as grid-based DEM. In a TIN model, the sample points are connected by lines to form triangles and within each triangle the surface is usually represented by a plane. By using triangles, it is ensured that each piece of the mosaic surface will fit with its neighboring pieces. Thus, the surface will be continuous, as each triangle's surface would be defined by the elevations of the three corner points.

Polygons that are more complex can also be used as mosaic tiles, but they can always be broken down into triangles. For vector GIS, TIN can be seen as polygons having attributes of slope, aspect, and area with three vertices having elevation attributes and three edges with slope and direction attributes. The TIN model is attractive because of its simplicity and economy. Certain types of terrain are very effectively divided into triangles with plane facets (e.g., fluvially eroded landscapes). Other landscapes, however, are not well represented by flat triangles (e.g., glaciated landscapes). Triangles work best in areas with sharp breaks in slope, where TIN edges can be aligned with breaks (e.g., along ridges or channels).

TIN Data Structure

Advantages:
1. Ability to describe the surface at different levels of resolution;
2. Efficiency in storing data;
3. TINS can include the highest/lowest points;
4. Can increase sampling in zones of high relief;
5. Is suitable for trend surfaces.

Disadvantage:
In many cases require visual inspection and manual control of the network.

1.4.4 Elements of a TIN

A TIN data model is composed of nodes, edges, triangles, and topology.

Nodes. Nodes are the fundamental building blocks of the TIN. The nodes originate from the points and arc vertices contained in the input data sources. Every node is incorporated in the TIN triangulation. Every node in the TIN surface model must have a z value.

Edges. Every node is joined with its nearest neighbors by edges to form triangles that satisfy the Delaunay criterion. Each edge has two nodes, but a node may have two or more edges. Because edges have a node with a z value at each end, it is possible to calculate a slope along the edge from one node to the other.

Triangles. Each triangular facet describes the behavior of a portion of the TIN's surface. The x, y, z coordinate values of a triangle's three nodes can be used to derive information about the facet, such as slope, aspect, surface area, and surface length. Considering the entire set of triangles as a whole, it is possible to derive additional information about the surface including volume, surface profiles, visibility analysis, and surface views.

Because each facet summarizes a certain surface behavior, it is important to ensure that the sample points are selected adaptively to give the best possible surface fit. A TIN surface model can yield poor results if important regions of the surface are undersampled.

Topology. The topological structure of a TIN is defined by maintaining information that defines each triangle's nodes, edge numbers and type, and adjacency to other triangles. For each triangle, TIN records (see Figure 1.7):

1. The triangle number;
2. The numbers of each adjacent triangle;
3. The three nodes defining the triangle;
4. The x and y coordinates of each node;
5. The surface z value of each node.

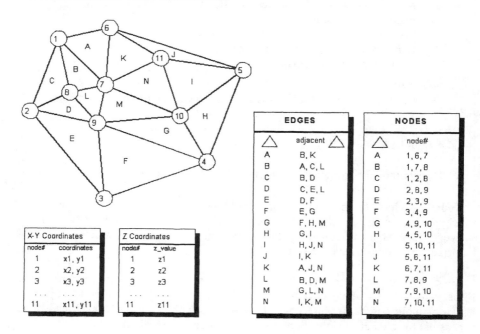

Figure 1.7 The structure of a TIN. Data is stored in a set of tables that retain the coordinate values as well as the spatial relations of the facets. (After Laurini and Thompson, 1992.)

1.4.5 Comparison of Grid-Based and Triangulation-Based DTMs

Among these DTM different data structures, grid-based and triangulation-based terrain modeling techniques are the most commonly used. In order to obtain a high-quality representation of the earth surface, several requirements have to be fulfilled (Kraus, 2000):

1. *Smoothing:* Random errors are contained in the original data. In order to reduce their influence on the resulting terrain model, filtering methods are required in the generation process.
2. *Geomorphological correctness:* Geomorphological information such as information on break lines or mountain peaks should be considered.

3. *Variable point density:* In order to reduce the amount of disk and memory space required for DTMs, methods allowing for variable point density (variable facet size) are required. In regions of high terrain curvature, a dense distribution of data points is required whereas flat and smooth regions can be modeled by a small number of large facets.
4. *Robustness:* The generation process should be robust with respect to the spatial distribution of the original data points.

Table 1.1 compares the grid and TIN data structures with respect to the above four criteria. Note that even though triangulation is based on 3-D algorithms, there are restrictions in the applicability due to the problems with the definition of the ordering criterion. With respect to modeling of buildings, both methods can be applied with certain drawbacks. Modeling buildings (without overhangs!) by grid-based models is possible if two break lines of different heights are introduced at the building outlines (one corresponding to the building outlines on the floor and the other one to the roofs; the latter has to be displaced slightly from the first one so that it is "inside" the outline on the floor, and the walls have to be slightly tilted so that no overhangs occur) and if break lines are used to model internal ridges. However, using grid-based modeling, a building cannot be addressed as an individual object in a data base. Modeling buildings by triangulation techniques is a special case of boundary representation. It can be accomplished, but not automatically, just based on a point cloud as input (Rottensteiner, 2001).

Table 1.1

A Comparison of Grid- and Triangulation-Based DTMs

	Grid DTM	*Triangulation*
Smoothing	Due to the generation algorithm based on least squares adjustment, grid-based methods perform smoothing.	Difficult to achieve because the original data points are used.
Geomorphology	Break lines can be considered.	Break lines can be considered.
Point density	Fixed due to the matrix structure. Usually, a trade-off between storage capacities and precision is sought.	Variable as the original data points are used.
Robustness	Robust estimation procedures can be applied. Problems appear with inhomogeneous point distributions.	Problems due to the nonuniqueness of the ordering criterion.
Applicability	Restrictions due to the 2.5-D characteristics. Simple algorithms can be found for many tasks.	More general than grid-based methods but also restricted. More difficult algorithms are required.

Source: Rottensteiner (2001).

1.5 GLOBAL ELEVATIONS DATA SOURCES

The worldwide concern over the widely perceived need to move towards greater sustainability has been reflected in the formulation of a number of agreements, many of which call for an increase in systematic observations of the Earth to increase our understanding of its processes. Examining the specific goals and objectives behind the establishment of these agreements, we find a real need at the most fundamental level to measure, map, monitor, and model baseline data in order to better manage our economic, social, and environmental resource base. Within the context of the International Steering Committee for Global Mapping (ISCGM), global data is defined as that which meets certain criteria: covers the entire globe; is at a scale of 1:1,000,000 or a resolution of 1 km; and includes specific core data layers (ISCGM, 1996).

Many global elevation data have been developed over the last decade and in fact there are global elevation data that exceed the ISCGM requirements in terms of spatial resolution. Global elevation data are suitable for many regional and continental applications, including climate modeling, continental-scale land cover mapping, extraction of drainage features for hydrologic modeling, and geometric and atmospheric correction of medium- and coarse-resolution satellite image data. The following will provide details about some of the current global elevation data. They include the following.

1. The Global Land One-km Base Elevation (GLOBE): GLOBE is an international effort to create a global digital elevation model (DEM) on a nominal 1-km grid and estimated absolute vertical accuracy ranging from 10 to 500 m at 90% linear error. GLOBE is designed, openly peer-reviewed, implemented, and documented while being coordinated by a global consortium of scientists and organizations. It is an activity of the Committee on Earth Observation Satellites (CEOS) and the International Geosphere-Biosphere Programme's Data and Information System (IGBP-DIS). GLOBE data came from 11 sources, via 18 combinations of source processing. Source data include satellite imagery, aerial photography, satellite altimetry, cadastral survey data, and hardcopy topographic maps converted to digital format. The source data were converted to 16-bit binary raster grids by a variety of techniques, including stereo profiling, image pattern recognition, contour-to-grid, and point-to-grid surface generation.

GLOBE data set is covering 180° west to 180° east longitude and 90° north to 90° south latitude. The horizontal grid spacing is 30 arc-seconds in latitude and longitude, resulting in dimensions of 21,600 rows and 43,200 columns. At the equator, a degree of latitude is about 111 km. GLOBE has 120 values per degree, giving GLOBE slightly better than 1 km gridding at the equator, and progressively finer longitudinally toward the poles. The horizontal coordinate system is seconds of latitude and longitude referenced to World Geodetic System 84 (WGS84). The vertical units represent elevation in meters above mean sea level. The elevation values range from -407m to 8,752m on land. In GLOBE Version 1.0, ocean areas have been masked as "no data" and have been assigned a value of -500.

2. Digital Terrain Elevation Database (DTED): DTED is a provided by the National Imagery and Mapping Agency (NIMA), with a horizontal grid spacing of 30 arc-seconds (approximately 1 km). The DTED horizontal datum is the World Geodetic System (WGS 84) and the vertical datum is the mean sea level (MSL) as determined by the Earth Gravitational Model (EGM) 1996. The data available to the public is called Level 0 and has a 30 arc-second spacing. Other higher resolution data called Level 1 (horizontal accuracy of 90% circular error WGS $\leq$ 50m and vertical accuracy of 90% linear error mean sea level (MSL) $\leq$ 30m) and Level 2 (horizontal accuracy of 90% circular error WGS $\leq$ 23m and vertical accuracy of 90% linear error MSL $\leq$ 18m) are not available to the public. Data density depends on the level produced. DTED Level 1 post spacing is 3 arc-seconds (approximately 100m), while DTED Level 2 post spacing is 1 arc-second (approximately 30m).

3. GTOPO30: GTOPO30 is a global digital elevation model (DEM) with a horizontal grid spacing of 30 arc-seconds (approximately 1 km). GTOPO30 was developed to meet the needs of the geospatial data user community for regional and continental scale topographic data. Completed in 1997, the data set was developed at the U.S. Geological Survey's EROS Data Center over a 3-year period with support from several national and international cooperators. It was derived from several raster and vector sources of topographic information which include Digital Terrain Elevation Data, Digital Chart of the World, USGS Digital Elevation Models, Army Map Service Maps, International Map of the World, Peru Map, New Zealand DEM, and the Antarctic Digital Database. The absolute vertical accuracy of GTOPO30 varies by location according to the source data. Generally, the areas derived from the raster source data have higher accuracy than those derived from the vector source data. The full resolution 3 arc-second DTED and USGS DEMs have a vertical accuracy of + or - 30m linear error at the 90% confidence level.

4. Shuttle Radar Topography Mission (SRTM): The SRTM is a joint project between the National Geospatial-Intelligence Agency (NGA) and the National Aeronautics and Space Administration (NASA). The SRTM program is a manned space-borne vehicle that has been fitted with a state-of-the-art radar system mounted within the bay of a space shuttle vehicle. The advantage of using radar versus other means is that radar works through all weather and functions day or night. It has the capability to image through clouds with minimal to no impact. The data used to produce the DEM is collected using a technique known as single-pass radar interferometery (see Chapter 2 for more details).

The driving purpose of this project is to provide detailed, accurate, and high-resolution global digital terrain elevation models of the earth (all land areas between 60° north and 56° south latitude), with data points located every 1 arc-second (approximately 30m) on a latitude/longitude grid. Some of the data that has been released are 3 arc-second (approximately 90m) with 16m absolute vertical height accuracy (at 90% confidence), 10m relative vertical height accuracy, and 20m absolute horizontal circular accuracy.

References

ESRI, ARC/INFO, *Surface Modeling with TIN™ User's Guide*, 1992.

International Steering Committee for Global Mapping (ISCGM), *Report of the Second Meeting of the International Steering Committee for Global Mapping*, Santa Barbara, CA, November 16, 1996.

Kraus, K., *Photogrammetrie Band 3. Topographische Informationssysteme*, Köln, Germany, Dümmler Verlag: 2000.

Laurini, R., and Thompson, D., *Fundamentals of Spatial Information Systems*, New York: Academic Press, 1992.

Miller, C., and LaFlamme, R.A., "The digital terrain modeling-theory and applications," *Photogrammetric Engineering*, 24(3): 433-442, 1958.

Petrie, G., and Kennie, T.J.M., *Terrain Modeling in Surveying and Civil Engineering*, New York: McGraw-Hill, 1990.

Peucker, T., Fowler, R., Little, J., and Mark, D., "The triangulated irregular network," *Proceedings of the Digital Terrain Models Symposium*, St. Louis, MO, 1978.

Rottensteiner, F., "Semi-automatic extraction of buildings based on hybrid adjustment using 3D surface models and management of building data in a TIS," Ph.D. thesis, Vienna University of Technology, Faculty of Science and Informatics, 2001.

Chapter 2

DTM Generation

2.1 INTRODUCTION

Traditionally, DTMs have been derived from ground surveys, photogrammetric techniques using film-based cameras and photogrammetric plotters (in its analog or analytical form), and from topographic data using contour data, spot heights, hydrology, and boundaries (shore line, state, tile). With the increasing demand for high-accuracy elevation and terrain information, new technologies have been developed to satisfy this demand. The last two decades have witnessed a number of technological developments that have changed the way we derive and update terrain models. These advances include the systems that collect the terrain data as well as the hardware and software utilized to manipulate, manage, and analyze it. Traditional methods of generating terrain models, including manual photogrammetry and field surveying, are rapidly leaving the marketplace to the new systems that combine data from remote sensing sensors (such as digital cameras, laser, and microwave remote sensing) with direct georeferencing navigation sensors. Combining the advances in remote sensing sensors and direct georeferencing has not only increased the efficiency of DTM generation considerably, but has also resulted in greater flexibility and lower cost. This chapter will review the four most popular approaches for generating terrain data: cartographic data, optical, laser, and microwave remote sensing.

2.2 CARTOGRAPHIC DATA SOURCES

Most of the DTMs currently available have been interpolated from contours by sampling designs and computer algorithms that add artifacts and other distortions inherent in the processing (Shortridge, 2001). This analog data may be digitized through manual digitization, semiautomated line following, or by means of automatic raster scanning and vectorization. The latter method is predominant for

19

large data collection projects. In principle, automated raster to vector conversion performs a similar task to manual digitizing using a digitizing tablet. However, there are many factors that can affect the accuracy of the digitizing process, such as resolution, consistency, stability, and others. With high-resolution scanned contour maps, raster to vector (R2V) methods track contour lines to the resolution of the scanned map. Scanning in fact deletes much of the human intervention and usually produces more accurate digital contours after editing (which is an unavoidable step in both manual digitizing and scanning) than manual digitizing.

The process of converting a paper contour map into vector format includes the following steps:

1.   Scanning the map;
2.   Removing noise;
3.   Detection, binarization, and skeletonization of the contours;
4.   Vectorization of the contours (contour following).

2.2.1 Scanning Maps

The scanning process converts the analog (paper) map into raster (digital) format. Scanning is an effective means of automating DTM data generation from contour maps. Recent developments in scanning hardware and R2V conversion software have made map scanning a viable tool for automated DTM generation from these maps. Automatic R2V technology encompasses a range of capabilities, from simple line generation to more complex object- or pattern-oriented recognition. In its simplest form, the process of tracing contours over raster is the most fundamental conversion approach.

The quality of contour R2V conversion depends largely on the quality of the source image that is affected by many factors including the scanner optical characteristics, cleanliness and age of the source map, scanning resolution, whether it is color or black/white, and others. Image processing techniques such as noise removal, removal of map background (especially for old maps with blue background), color separation for color maps, and distortion correction play a major role in determining the quality of the final product.

One of the major challenges in DTM generation from scanned maps is automatic text recognition of the contour elevation. To reliably recognize text labels in maps, the first step is to separate them from lines, also known as the text segmentation step. Once separated, text recognition engines are applied to identify the text and convert them to computer readable ASCII code, or unicode for other languages such as Chinese and Japanese. Conventional text recognition (OCR) technologies have not worked well for recognizing text in maps and drawings largely due to the variety of fonts, sizes, and orientations used. International languages add more difficulty to the problem (Wu, 2000).

In any scanning software, the selection of appropriate dots per inch (DPI) for the scan is in essence the determining factor of how many dots per inch the scanner will record. The more DPI, the more bits (binary digits) are needed, and the greater the spatial resolution and file size of the image. This usually raises the question: To what DPI should a map be scanned? As a rule of thumb, at least 2 to 4 pixels are required to represent a feature in the scanned map. Therefore the DPI should be selected to produce the scanned contours with a minimum of 2 to 4 pixels. For example, a contour line of 0.25mm thickness will be equivalent to 10 pixels at 1,000 DPI (one dot in μm for 1,000 DPI = 2.54 cm × 10,000/DPI = 25 μm) and 5 pixels at 500 DPI. In general, the minimum number of DPI to provide (n) pixels for a contour of thickness (s) can be written as, $n = \{s$ (in millimeters)$\}$/dot size = $\{s$ (in millimeters) × DPI$\}$/24.5 mm, from which we can estimate the DPI = 25.4 n/s.

2.2.2 Removing Noise

Noise is commonly present in any scanned map. Noise can originate as a result of a poor sampling process or from the original map. The purpose of removing the noise from the scanned map is to reduce unwanted noise before detecting, binarizing, and vectorizing the contours. Noise removal is usually accomplished through the implementation of spatial domain smoothing techniques in local neighborhoods of the scanned map. The following shows two examples of local neighborhood operators for noise removal:

1.   Median filter: The process of median filtering consists of sorting the intensity values in ascending or descending order, then choosing the median or middle value as the new intensity value of the pixel we are working with as shown here:

$$\begin{bmatrix} 93 & 100 & 98 \\ 105 & 25 & 95 \\ 107 & 102 & 100 \end{bmatrix} \rightarrow 25 \quad 93 \quad 95 \quad 98 \quad 100 \quad 100 \quad 102 \quad 105 \quad 107 \rightarrow \begin{bmatrix} 93 & 100 & 98 \\ 105 & 100 & 95 \\ 107 & 102 & 100 \end{bmatrix}$$

If one pixel in the neighborhood is dramatically different (noise) from the rest, a median filter would completely remove it. Another attribute of a median filter is that it does preserve sharpness of an image.

2. Neighborhood averaging: This technique is a simple, intuitive, and easy-to-implement method for reducing noise in images. The idea of neighborhood averaging is simply to replace each pixel value in an image with the mean (average) value of the pixel and its neighbors. This has the effect of eliminating pixel values that are unrepresentative of their surroundings. Mean filtering is usually thought of as a convolution filter. Like other convolutions techniques, it is based around a kernel that represents the shape and size of the neighborhood to be sampled when calculating the mean. Often a 3×3 square kernel is used. However, larger kernels (e.g., 5×5 squares) can be used for more severe smoothing. The

following shows an example of averaging the 3×3 neighborhood used in the median filter example, where the averaged value (92) will replace the central pixel value (25) in the averaged images:

$$\frac{1}{9} * \begin{bmatrix} 1 & 1 & 1 \\ 1 & 1 & 1 \\ 1 & 1 & 1 \end{bmatrix} * \begin{bmatrix} 93 & 100 & 98 \\ 105 & 25 & 95 \\ 107 & 102 & 100 \end{bmatrix} = 92$$

2.2.3 Contour Detection, Binarization, and Skeletonization

2.2.3.1 Contour Detection

Scanned contours are linear features. Their boundaries therefore represent edges (the transition between the contours and background). An edge in the image-processing sense is a discontinuity in the two-dimensional grayscale functions. Contour detection refers to the processes that examine the scanned map for discontinuities in the gray levels. An edge is defined as the boundary between the contour line and the background.

For continuous functions, differentiation is used to determine the rate of change in a function. The rate of change in the gray level intensity of an area of the image space constitutes an edge. The approach taken for most edge detection techniques is to find some sort of local derivative to determine the change in the gray level intensity. It is important that a scanned map has been properly pre-processed before progressing with edge detection. Preprocessing consists of removing any distracting data from the scanned map as described in the previous section. The goal of edge extraction is to remove the background from the scanned map. This is usually achieved by differentiation. The most commonly used method of differentiation in image-processing applications is the gradient (Gonzalez, 1987). Given a grayscale function $g(x,y)$, the gradient of g at coordinates (x, y) is defined as the vector:

$$G[g(x, y)] = \begin{bmatrix} \partial g / \partial x \\ \partial g / \partial y \end{bmatrix} \tag{2.1}$$

and the magnitude of $G[g(x,y)]$ can be given and simply written as:

$$G[g(x, y)] = [(\partial g / \partial x)^2 + (\partial g / \partial y)^2]^{1/2} \tag{2.2}$$

which represents the maximum rate of increase of $g(x, y)$ per unit distance in the direction of G. For the discrete gradient digital images, some gradient filters can be used to find the edge pixels. There exist a number of estimation procedures for determining the respective components of the gradient. One of the most commonly used convolution kernels for edge detection and enhancement of scanned images was given by (Chavez, 1975).

The kernel is specified as:

$$\text{Chavez kernel} = \frac{1}{9} * \begin{bmatrix} -1 & -1 & -1 \\ -1 & 17 & -1 \\ -1 & -1 & -1 \end{bmatrix} \tag{2.3}$$

For a particular pixel location, a low-pass filter may be used to evaluate the average value in a 3×3 window:

$$\text{AVG} = \frac{1}{9} * \begin{bmatrix} 1 & 1 & 1 \\ 1 & 1 & 1 \\ 1 & 1 & 1 \end{bmatrix} \tag{2.4}$$

The high-frequency (*HF*) component in any given pixel will then be given by:

$$\text{HF} = \text{pixel value - average value} \tag{2.5}$$

Represented in terms of a convolution kernel, this would be

$$\text{HF} = \begin{bmatrix} 0 & 0 & 0 \\ 0 & 1 & 0 \\ 0 & 0 & 0 \end{bmatrix} - \frac{1}{9} * \begin{bmatrix} 1 & 1 & 1 \\ 1 & 1 & 1 \\ 1 & 1 & 1 \end{bmatrix} \quad \text{HF} = \frac{1}{9} * \begin{bmatrix} -1 & -1 & -1 \\ -1 & 8 & -1 \\ -1 & -1 & -1 \end{bmatrix} \tag{2.6}$$

By adding the high frequency part, *HF*, back to the original pixel, a high-frequency enhancement will be achieved:

$$\text{New value} = \text{pixel value} + \text{HF} \tag{2.7}$$

$$\text{New value} = \begin{bmatrix} 0 & 0 & 0 \\ 0 & 1 & 0 \\ 0 & 0 & 0 \end{bmatrix} + \frac{1}{9} * \begin{bmatrix} -1 & -1 & -1 \\ -1 & 8 & -1 \\ -1 & 1 & -1 \end{bmatrix} \tag{2.8a}$$

or

$$\text{New value} = \frac{1}{9} * \begin{bmatrix} -1 & -1 & -1 \\ -1 & 17 & -1 \\ -1 & -1 & -1 \end{bmatrix} \qquad (2.8b)$$

Other methods used for edge detection include Sobel, Roberts, and Laplacian operators. For more details about these filters the reader is advised to consult (Gonzalez, 1987) or any other digital image processing book.

2.2.3.2 Binarization

The input grayscale image has to be binarized before the skeletonization stage. At this point of processing, we are dealing with scanned maps that have been adequately enhanced, smoothed, and the contours have been detected as edges. The purpose of the binarization procedure is to segment the scanned map into background and contour pixels

The binarization of grayscale scanned maps can be performed by thresholding with some threshold T. If the pixel value $f(x, y)$ is less than the threshold T, the pixel belongs to the background and the thresholded pixel value will be 0. Otherwise the pixel belongs to a contour and the value 1 will be assigned to the pixel. Common representation of the segmented map $g(x, y)$ using a threshold value T is given by the following: For an input scanned map $f(x, y)$, the output segmented scanned map $g(x, y)$ using a threshold value T is given by:

$$g(x, y) = \begin{cases} 1 \xrightarrow{\;if\;} f(x, y) \geq T \\ 0 \longrightarrow \text{otherwise} \end{cases} \qquad (2.9)$$

There are basically two classes of binarization (thresholding) techniques: global and adaptive. Global methods binarize the entire image using a single threshold. A simple way to select a global threshold automatically is to use the value at the valley of the intensity histogram of the image; this assumes that there are two peaks in the histogram, one corresponding to the contour, the other to the background. In practice, because of nonuniformity of the background in the scanned map, global thresholding can provide poor results.

While the global thresholding operator uses a global threshold for all pixels, adaptive thresholding changes the threshold dynamically over the image. In this case, the scanned map is divided into rectangular blocks and each of them is processed with an adaptive threshold defined by the statistical properties of the block. The assumption behind the use of adaptive thresholding is that smaller image regions are more likely to have approximately uniform illumination during the scanning process, thus being more suitable for thresholding.

There are a number of techniques for calculating the proper thresholding values; see for example Weszka (1878), and Pal (1993), Glasbey (1993), and

Jawahr et al. (2000). An approach to finding the local threshold is to statistically examine the intensity values of the local neighborhood of each pixel. The statistic that is most appropriate depends largely on the input scanned map. Simple and fast functions include the *mean* of the *local* intensity distribution, or the mean of the minimum and maximum values.

The size of the neighborhood has to be large enough to cover sufficient foreground and background pixels, otherwise a poor threshold is chosen. On the other hand, choosing regions that are too large can violate the assumption of approximately uniform illumination.

2.2.3.3 Skeletonization

After the edge detection and binarization procedure, the detected edge pixels are recorded. In this case, all the pixels in the area of interest are divided into two classes: contour (edge) pixels and background pixels. This processing is called binarization. In order to distinguish between edge pixels and nonedge pixels, a predefined global threshold is used. The result of the gradient filtering and the binarization are edges wider than one pixel. The required final position of the edge lies approximately in the middle of this wider edge (Heipke et al., 1994). To extract the center position of the edge, some algorithms should be applied. This is known as skeleton processing.

For the purpose of skeletonizing the edge pixels, one way is to check the small area containing the candidate edge pixel to decide whether the represented pixel belongs to the skeleton. In this situation, the template method can be applied to solve the problem. The simple and popular template is based on a square array of image. Normally the size may be 3×3, 5×5, 7×7, and so forth. As the template size increases, the number of different combinations dramatically increases and the computation time will be unacceptable.

Here, an algorithm proposed by Kreifelts (1977) is implemented. This approach works in a 3×3 array of graylevel values. The center pixel is the represented pixel and it is also a contour pixel. Comparing different 3×3 template arrays (see Figure 2.1) the examined array will be classified either as a skeleton pixel or a nonskeleton pixel. The nonskeleton pixel will be removed. This is a process by which a contour binary image is reduced to thin (one pixel wide) lines. The result is a skeleton line that is close to a medial line.

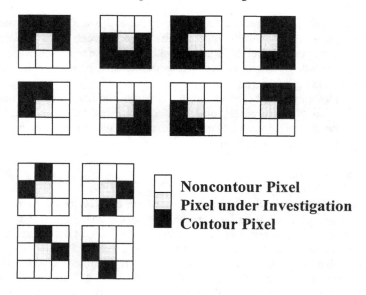

Figure 2.1 Template matrices for skeletonization of scanned contour lines.

Figure 2.2 shows part of the binary image after edge detection and binarization. The pixel under investigation contains a contour pixel that does not belong to a skeletonized contour line. Therefore, the center pixel should be changed to a background pixel. The processing result is shown in the one on the right.

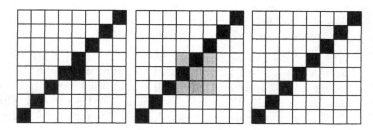

Figure 2.2 Description of the detailed skeletonization procedure.

Computational efficiency is one of the most critical factors in the whole skeletonization procedure. When comparing the 3×3 pixel array with the template matrices, it is necessary to pass over all the templates one by one. These templates can be considered as different conditions. Therefore, it would require a large computation time. To reduce the computation time, a binary coding method of comparison is usually applied. Each position of the eight neighboring pixels in a 3×3 matrix has an index number from 0 to 7 (see Figure 2.3).

Figure 2.3 Ordering of neighborhood pixels.

If the pixel is a contour pixel, it is assigned a 1; otherwise, it is assigned a 0 (as described in the binarization section). Therefore, we can form an 8-bit binary code from the neighboring pixels. On the other hand, a one-dimensional array with (n) elements is allocated and also assigned values, where (n) is the number of template matrices. The assigned values are either 1 or 0. The value 1 means that the center pixel is a skeleton pixel and 0 means it is not. Let's take the following matrix as an example:

which corresponds to with the numbering convention .

Its binary code will be [010000010] which corresponds to [66] in digital. When the 3×3 image array is checked, its binary code is first formed as i. If the binary number i is converted into digital format and it equals the digital number 66, that means that the corresponding center pixel is not a skeleton pixel and should be changed to a background pixel.

2.2.4 Contour Following

After edge skeletonization, thin contour lines with one pixel width in the area of interest are obtained. For extracting the whole contour, we need to trace pixels and obtain their positions. This process is known as contour following or vectorization. The vectorization process is usually done in semiautomatic mode, where the operator provides the initial points. The initial direction can either be given by the operator or be determined through the automatic search procedure. In the latter case, the initial direction is actually approximated as $0°$ (i.e., pointing upward in the scanned map).

The user usually defines the number of search directions. For example, if the user defined the initial direction as $0°$ and the number of directional matrices as 13, then the direction will cover the range from $-90°$ to $+90°$ with one for every

15°. Figure 2.4 displays the directional matrices from 0° to +90° (by symmetry, the remainder of the matrices can easily be determined).

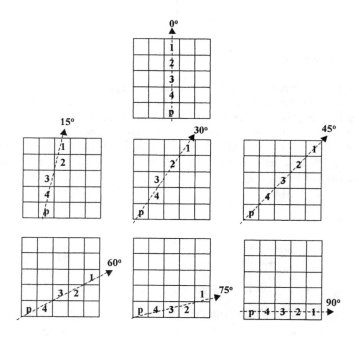

Figure 2.4 Directional matrices.

In each matrix shown in Figure 2.4, *p* defines the current position of the contour in processing. The searching action is performed in the order of the numbered pixels until the edge pixel is found in the matrix or changes to an adjacent directional matrix and continues. When a contour pixel is found, it is used as the new starting point by the algorithm and the procedure is repeated. If no edge pixel is found through all the directional matrices, a major gap might exist. Then, a larger directional matrix is applied to bridge it (Heipke et al., 1994) or the user has to provide another starting point. Xin (1995) uses directional matrices to vectorize road centerlines from close range images.

2.2.5 Generating Grids from Contour Data

The procedures necessary to convert the raster-scanned contour maps into vectors have been outlined in the previous sections. The next step after vectorizing the contours is to tag the elevation of the contour to each corresponding contour line. This process is usually done manually. The next step is then to create grids from these contours. Leberl and Olson (1982) described a height gridding method procedure which they term a "sequential steepest slope algorithm" (see Figure 2.5). In their method, a search is made along each of four lines passing through the required grid node and oriented along the grid directions (*VV* and *HH*) and their bisectors (*UU* and *GG*). The intersection of each of the eight directions with the nearest contours is established, and the slope of each of the four lines is calculated. The line with the steepest slope is then selected and the value of the elevation of the grid node is calculated using linear interpolation along this line. In the example shown in Figure 2.5, search line *GG* is the steepest and the height of the grid node *P* is derived as follows:

$$Z_p = \frac{H_1 - H_5}{d_{15}} d_{p5} + H_5 \qquad (2.10)$$

The implementation of the procedures usually avoids the calculation slope of lines if the end points intersect the same contour (lines *HH* and *UU* in this example).

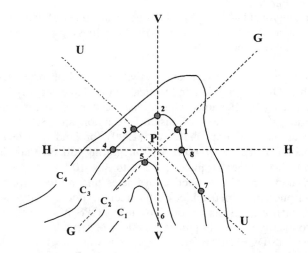

Figure 2.5 Sequential steepest slope algorithm showing cross sections HH, VV, UU, and GG and the intersection points 1-8 on the contour. (After Leberl and Olson, 1982.)

2.3 DEVELOPMENT OF GEOREFERENCING TECHNOLOGY

Direct georeferencing is the determination of time-variable position and orientation parameters for a remote sensing sensor. The most common technologies used for this purpose today are satellite positioning by GPS and inertial navigation using an inertial measuring unit (IMU). The fact that the method is in principle independent of available GCP has obvious economic advantages, especially in areas with poor or sparse control. The principle of direct georeferencing results in a number of systems that have been thoroughly tested and has resulted in georeferencing parameters of high accuracy through the use of integrated differential GPS and IMU systems in post-mission. Although each technology can in principle determine both position and orientation, they are usually integrated in such a way that the GPS receiver is the main position sensor, while the IMU is the main orientation sensor. The orientation accuracy of an IMU is largely determined by the gyro drift rates, typically described by a bias (constant drift rate), the short-term bias stability, and the angle random walk. Typically, four classes of gyros are distinguished according to their constant drift rate, namely (Schwarz and El-Sheimy, 2004):

1. Strategic gyros (0.0005–0.0010 deg/h) achieving an accuracy of degrees per month;
2. Navigation-grade gyros (0.002–0.01 deg/h) achieving an accuracy of degrees per week;
3. Tactical gyros (1–10 deg/h) achieving an accuracy of degrees per hour;
4. Low-accuracy gyros (100–10,000 deg/h) achieving an accuracy of degrees per second.

Since strategic gyros are rare and costly, only the last three classes will be discussed further. Only navigation and tactical grades have actually been used for direct georeferencing applications. Operational testing of direct georeferencing started in the early 1990s; see, for instance, Cannon and Schwarz (1990) for airborne applications, and Lapucha et al. (1990) for land-vehicle applications. These early experiments were done by integrating differential GPS with a navigation-grade IMU (accelerometer bias: 2 to 3 $\times$ 10^{-4} m/s^2, gyro bias: 0.003 deg/h) and by including the derived coordinates and attitude (pitch, roll, and azimuth) into a photogrammetric block adjustment. Although GPS was not fully operational at that time, results obtained by using GPS in differential kinematic mode were promising enough to pursue this development. As GPS became fully operational, the INS/DGPS georeferencing system was integrated with a number of different imaging sensors. Among them were the CASI sensor manufactured by ITRES Research Ltd. [see Cosandier et al. (1993)]; the MEIS of the Canada Centre for Remote Sensing; and a set of CCD cameras [see El-Sheimy and Schwarz (1993)]. Thus, by the end of 1993, experimental systems for mobile mapping existed for both airborne and land vehicles. A more detailed overview of the state of the art at that time is given in Schwarz et al. (1993). The evolution of georeferencing technology during the past decade was due to the ongoing refinement and miniaturization of GPS-receiver hardware and the use of low- and medium-cost IMUs that became available in the mid-1990s. Only the latter development will be briefly discussed here.

The inertial systems used in INS/GPS integration in the early 1990s were predominantly navigation-grade systems, typically strapdown systems of the ring-laser type. When integrated with DGPS, they provided position and attitude accuracies sufficient for all accuracy classes envisaged at that time. These systems came, however, with a considerable price tag (about $130,000 at that time). With the rapidly falling cost of GPS-receiver technology, the INS became the most expensive component of the georeferencing system. Since navigation-grade accuracy was not required for the bulk of the low- and medium-accuracy applications, the emergence of low-cost IMU in the mid-1990s provided a solution to this problem. These systems came as an assembly of solid-state inertial sensors with analog readouts and a postcompensation accuracy of about 10 deg/h for gyro drifts and about 10^{-2} m/s^2 for accelerometer biases. Prices ranged between $10,000 and $20,000 and the user had to add the A/D portion and the navigation software. Systems of this kind were obviously not suited as stand-alone navigation systems because of their rapid position error accumulation. However, when provided with high-rate position and velocity updates from differential GPS (1s pseudorange solutions), the error growth could be kept in bounds and the position and attitude results from the integrated solution were suitable for low- and medium-accuracy applications. For details on system design and performance, see Bäumker and Matissek (1992), and Lipman (1992), among others.

With the rapid improvement of fiber-optic gyro performance, the sensor accuracy of a number of these systems has improved by about an order of magnitude (1 deg/h and 10^{-1} m/s^2) in the past 5 years. Typical costs are about $30,000. Beside the increased accuracy, these systems are more user friendly and offer a number of interesting options. When integrated with a DGPS phase solution the resulting position and attitude are close to what is required for the high-accuracy class of applications. When aiming at highest possible accuracy, these systems are usually equipped with a dual-antenna GPS, aligned with the forward direction of the vehicle. This arrangement provides regular azimuth updates to the integrated solution and bounds the azimuth drift. This is of particular importance for flights flown at constant velocity along straight lines, as is the case for photogrammetric blocks (Schwarz and El-Sheimy, 2004).

2.4 DIRECT GEOREFERENCING: MATHEMATICAL MODELING

The formulation of the direct georeferencing mathematical model is rather straightforward. For details, see, for instance, Schwarz (2000). The standard implementation of this formula will, however, cause difficulties when low-accuracy gyros are used. The modifications necessary in this case will be discussed in this chapter. Figure 2.6 depicts airborne mobile mapping using a digital frame camera. The mathematical model is given in (2.11) for a camera system and will be used as the standard model in the following discussion. The terms in the equation are listed in Table 2.1.

$$\mathbf{r}_i^m = \mathbf{r}(t)_{GPS}^m + \mathbf{R}_b^m(t)(s_i \mathbf{R}_c^b \mathbf{r}_i^c + \mathbf{a}_{INS}^c - \mathbf{a}_{INS}^{GPS}) \qquad (2.11)$$

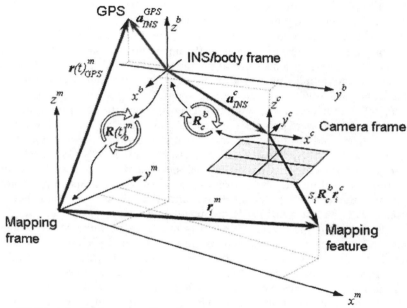

Figure 2.6 Principle of airborne georeferencing.

Implementation of this formula requires inertial and GPS measurements for the determination of the two time-dependent terms on the right-hand side of (2.11), as well as image coordinate measurements, in the c-frame, for the determination of the object point coordinate vector $\mathbf{r}_i^c$. The c-frame has its origin in the perspective center of the camera and its z-axis is defined by the vector between the perspective center and the principal point of the photograph. Its (x,y)-axes are defined in the plane of the photograph and are measured with respect to the principal point.

Table 2.1

Elements of the Georeferencing Formula

Variable	*Obtained from*
$\mathbf{r}_i^m$	The coordinate vector of point (*i*) in the mapping frame (m-frame) (3 unknowns)
$\mathbf{r}(t)_{GPS}^m$	The interpolated coordinate vector of GPS in the m-frame
s_i	A scale factor, determined by stereo techniques, laser scanners, or from DTM
$\mathbf{R}_b^m(t)$	The interpolated rotation matrix between the navigation sensor body frame (b-frame) and the m-frame
(t)	The time of exposure (i.e., the time of capturing the images), determined by synchronization
$\mathbf{R}_c^b$	The differential rotation (boresight) between the camera (c) frame and the INS body (b) frame, determined by calibration
$\mathbf{r}_i^c$	The coordinate vector of the point (*i*) in the c-frame (i.e., image coordinate)
$\mathbf{a}_{INS}^c$	The lever arm vector between INS center and camera principal point, determined by calibration
$\mathbf{a}_{INS}^{GPS}$	The lever arm vector between INS center and GPS antenna center, determined by calibration

The corresponding image vector is therefore of the form:

$$\mathbf{r}_i^c = \begin{pmatrix} x - x_p \\ y - y_p \\ -f \end{pmatrix} \tag{2.12}$$

where (x_p, y_p) are the principal point coordinates and f is the camera focal length.

If instead of a frame camera, either a pushbroom scanner or a LIDAR system (see Section 2.7) are modeled, the only change necessary is in the term r^c. It will have the form for:

$$\text{A pushbroom scanner} = \begin{pmatrix} 0 \\ y - y_p \\ -f \end{pmatrix} \tag{2.13}$$

A LIDAR scanner $= \begin{pmatrix} -d \cdot \sin \alpha \\ 0 \\ -d \cdot \cos \alpha \end{pmatrix}$ (2.14)

where d is the raw laser range (distance) in the laser frame, and α is the scanner angle from nadir along the y-axis of laser frame (see Figure 2.7).

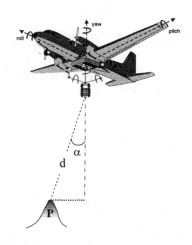

Figure 2.7 LIDAR scanner angle and distance measurement.

In addition, the misalignment matrix $\mathbf{R}_c^b$ and the offset vectors $\mathbf{a}_{INS}^c$ and $\mathbf{a}_{INS}^{GPS}$ have to be determined by calibration. A detailed discussion of implementation aspects can be found in El-Sheimy (1996) for land-vehicle applications and in Mostafa and Schwarz (2000) for airborne applications.

Using Newton's second law of motion in a rotating coordinate frame, the inertial sensor measurements, specific force f^b and angular velocity $\boldsymbol{\omega}_{ib}^b$, measured in the b-frame, can be transformed into an Earth-fixed frame, say the conventional terrestrial coordinate frame (e). The resulting system of differential equations is of the form

$$\begin{pmatrix} \dot{\mathbf{r}}^e \\ \dot{\mathbf{v}}^e \\ \dot{\mathbf{R}}_b^e \end{pmatrix} = \begin{pmatrix} \mathbf{v}^e \\ \mathbf{R}_b^e \mathbf{f}^b - 2\Omega_{ie}^e \mathbf{v}^e + \mathbf{g}^e \\ \mathbf{R}_b^e (\Omega_{ib}^b - \Omega_{ie}^b) \end{pmatrix}$$ (2.15)

where Ω_{ib}^b is the skew-symmetrical form of the angular velocity vector ω_{ib}^b, and the dot above a vector (bold lowercase) or a matrix (bold capital) indicates differentiation. Note that the m-frame in the georeferencing formula (2.11) has been replaced by the e-frame. This system is integrated to yield the parameters on the left-hand side, namely position, velocity, and the orthogonal rotation matrix $\mathbf{R}_b^e$ between the b-frame and the e-frame. The determination of this time-variable matrix is one of the central tasks of georeferencing.

The integration of the system of differential equations is started by initializing the parameters on the left-hand side of the equation. Traditionally, this is done in stationary mode. In that case, the initial velocity is zero, the initial position is obtained from GPS, and the initial orientation matrix is determined by an alignment procedure that makes use of accelerometer leveling and gyro compassing. The alignment is usually done in two steps. In the coarse alignment, simplified formulas are used to obtain pitch and roll from accelerometer measurements and azimuth from gyro measurements, within an accuracy of a few degrees. In the fine alignment, small-angle error formulas are used in a filtering scheme to obtain more accurate estimates of the parameters. The rotation matrix $\mathbf{R}_b^e$ can then be obtained from the estimated pitch, roll, and azimuth, or an equivalent parameterization, as for instance quaternions. This is the standard alignment procedure if sensor measurements from a navigation-grade inertial system are available. For medium- and low-accuracy systems, this method cannot be applied. Because of the unfavorable signal-to-noise ratio for ω_{ib}^b, the gyro compassing procedure will not converge. Thus, stationary alignment cannot be used with MEMS-based or other low-accuracy IMUs.

The alternative is in-motion alignment. This method has mainly been used in airborne applications, specifically for the in-air alignment of inertial systems. It is obviously dependent on using additional sensors, which in this case are GPS receiver outputs (i.e., position and/or velocity in the e-frame). In-motion alignment makes use of the fact that very accurate position information is available at a high data rate. It is therefore possible to determine accurate local-level velocities. By combining these velocities with the ones obtained from the strapdown IMU, the rotation matrix $\mathbf{R}_b^e$ can be determined.

When implementing this approach for medium- or low-accuracy IMUs, two difficulties have to be addressed. The first one is that either no azimuth information at all is available, or that it is derived from GPS velocities and is rather inaccurate. Thus, the standard small-angle error models cannot be used any more because the azimuth error can be large. By reformulating the velocity error equations, one can arrive at a set of equations that converges quickly for azimuth, even if the initial errors are large. Scherzinger (1994) has given a thorough discussion of the problem and its solutions, based on earlier work by Benson (1975). A land-vehicle application, using a MEMS-IMU integrated with GPS, is given in Shin and El-Sheimy (2004). Convergence is fast in this case, taking only about 50 seconds. This bodes well for airborne applications where, due to the higher velocities, a better signal-to-noise ratio should further improve the results.

The second difficulty is more fundamental. It has to do with the way in which the nonlinearity in equation system (2.15) and the corresponding GPS measurements are handled. The standard approach is to expand the errors in position, velocity, and orientation into a Taylor series and to truncate the series after the linear term. The error equations obtained in this way are then cast into state vector form. By adding the linearized GPS measurements to the model and by representing the state variable distribution by a Gaussian random variable, the extended Kalman filter (EKF) can be formulated. It is a standard tool in engineering that is frequently used when either the system model or the observation model are nonlinear. In an interesting paper Julier and Uhlmann (1996) have demonstrated that even in a seemingly innocuous situation — a road vehicle moving along a circle — the EKF does not handle the nonlinearity in an acceptable manner. After a quarter circle, the covariance propagation results in error ellipses that do not represent the actual situation. The authors show quite convincingly that this is due to the linearized covariance propagation. To rectify the situation, the authors propose to change this part of the EKF. They approximate the Gaussian distribution at a carefully chosen set of points and propagate this information through the nonlinear equations. In this way the transformed Gaussian distribution will reflect the nonlinearities of the system better. The new filter, called the unscented Kalman filter (UKF) by its authors, has received some attention during the past few years; see, for instance, Julier and Uhlmann (2002) and Crassidis and Markley (2003). A paper that combines in-motion alignment, large azimuth modeling, and UKF for MEMS /DGPS integration is Shin and El-Sheimy (2004). Although the UKF does not often show a substantial increase in accuracy, it seems to be more robust than the EKF in critical situations.

2.5 EXAMPLES OF DIRECT GEOREFERENCING SYSTEMS

Commercialization of the direct georeferencing systems for all application areas has been done by Applanix Corporation (now a subsidiary of Trimble, http://www.applanix.com) and iMAR GmbH (http://www.imar-navigation.de). The tables of results shown in the following are not based on a comprehensive analysis of published results. They are rather samples of results achieved with specific imaging systems and have been taken from company brochures and technical publications. The authors think that they are representative for the systems that have been discussed previously.

The accuracy specifications for the Applanix family of POS/AV (see Figure 2.8) airborne direct georeferencing systems are listed in Table 2.2 (Mostafa et al., 2000). The primary difference in system performance between the POS/AV 310 and POS/AV 510 systems is the orientation accuracy, which is directly a function of the IMU gyro drifts and noise characteristics. For example, the gyro drifts for the POS/AV 310 and POS/AV 510 systems are 0.5 deg/hr and 0.1 deg/hr, respectively, while the corresponding gyro noises are 0.15 and 0.02 deg/hr$^{-1/2}$, respectively.

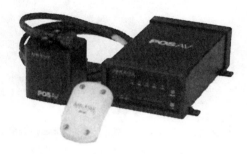

Figure 2.8 The Applanix POS-AV system. (Courtesy of Applanix Corporation.)

Table 2.2

Post-Processed POS/AVTM Navigation Parameter Accuracy

Parameter Accuracy (RMS)	Model 310	Model 410	Model 510
Position (m) [DGPS]	0.1 – 0.3	0.05 – 0.3 m	0.01 – 0.03
Velocity (m/s)	0.01	0.005	0.005
Roll / pitch (deg)	0.013	0.008	0.005
True heading (deg)	0.035	0.015	0.008
Gyro drift (deg/hr)	0.5	0.5	0.1
Noise (deg/sqrt(hr))	0.15	0.07	0.02

Source: Mostafa et al. (2000).

The accuracy specifications for the iMAR family of iNAV-FJI-AIRSURV (see Figure 2.9) for one of the iMAR IMUs airborne direct georeferencing systems are listed in Table 2.3.

Figure 2.9 The iMAR iNAV-RQH-003-AIRSURV system. (Courtesy of iMAR GmbH.)

Table 2.3

Post-Processed iNAV-xxx-AIRSURV Navigation Parameters Accuracy (Provided by iMAR GmbH)

Parameter Accuracy (RMS)	iNAV-FMS-AIRSURV	iNAV-RQH-003-AIRSURV	iNAV-FJI-0003-AIRSURV
Position (m) [DGPS]	0.1–0.3	0.05–0.3m	0.01–0.03[1]
Velocity (m/s)	0.01	0.003	< 0.002
Roll / pitch (deg)	0.025	0.005	< 0.002
True heading (deg)	0.06	< 0.01	0.005
Gyro drift (deg/hr)	0.75	0.003	0.001[2]
Gyro ARW (deg/sqrt(hr))	0.15	0.003	0.0003
Technology	FOG	RLG	FOG

[1]   Absolute position accuracy depends on environmental conditions and the distance between the target and the DGPS base station. On request iMAR offers a network solution using several reference stations. Also a multiantenna assembly is available to reduce remaining multipath (see also iMAR's iSMARTpos-3D).

[2]   Drift stability 0.001 deg/hr. Initial drift over temperature of iNAV-FJI-0003 is < 0.01 deg/hr, but is estimated by the integrated Kalman filter after 10 minutes of flight.

    The systems cover a wide range of performance of 0.75 to 0.001 deg/hr. The iNAV-RQH-AIRSURV uses ring laser gyro technology and is used where an excellent initial gyro bias over temperature is required (standard: 0.003 deg/hr, option: 0.002 deg/hr).

2.6 PHOTOGRAMMETRIC DATA CAPTURE

2.6.1 The Analog Photogrammetry Era

Photogrammetry has established itself as the main technique for obtaining precise 3-D measurements from stereo overlapped imagery. Conventional photogrammetry involves the use of expensive plotting equipment to mimic the stereo geometry (stereo model) at the moment of image exposure. This is accomplished through two steps: (1) the inner orientation that establishes the relationship between the camera space coordinates and the model space coordinates through the use of either fiducials or reseau marks, and (2) the relative orientation process that establishes the correct position and orientation of one photograph relative to the adjacent overlapping photograph. This is achieved by enforcing the condition that conjugate rays intersect. When relative orientation is complete the so-called photogrammetric model is established (Shears and Allan, 1996).

Once oriented, all residual y-parallax will have been eliminated, allowing the operator to view the model in stereo. The last step is to relate the model coordinate system with the ground coordinate system through the absolute orientation process. Absolute orientation is the process of solving the orientation elements, which relate an arbitrary system of coordinates (the model system in our case) to the reference system (the ground coordinate system). The absolute orientation is done analytically with the help of model and ground coordinates of a minimum number of two planimeteric and three height (i.e., with known X, Y, Z in both systems) control points.

Once the model has been absolutely oriented, it is possible to extract the DTM through different sampling techniques. It is possible to distinguish a number of different photogrammetric sampling techniques: regular sampling patterns, progressive sampling, selective sampling, and contouring. Each of these methods attempts to minimize the data collection effort, while at the same time maximizing the accuracy of the resulting DTM.

Regular sampling patterns (systematic): This is typically used in analytical plotters where regular patterns may be arranged as profiles or regular geometric shapes (square, rectangular, triangular). Since a fixed sampling distance is used, it is important to consider the determination of the optimal sampling interval. Location of the required grid node is preprogrammed and driven under computer control.

Progressive sampling: The density of sample points is adapted to the complexity of the terrain surface. The sampling progress is initiated by measuring a low-density grid. The accuracy of the sampled data is then analyzed. Wherever necessary, the sampling grid is recursively densified until the required accuracy level is reached. This results in a hierarchical sampling pattern.

Selective sampling (random sampling): This method can be combined with progressive sampling, called composite sampling. In this technique, selective sampling is used to capture abrupt topographic changes while progressive sampling yields the data for the rest of the terrain.

Contouring: Systematically measures contours over the whole area of the stereo model and outputs the measurements in the form of strings of digital coordinates. Method may be supplemented by spot heights measured along terrain break lines.

This process has some fundamental drawbacks when compared to digital techniques. First, it is all based on very specialized hardware; namely analog and analytical plotters. Both plotters however are designed to carry out these single specific tasks and cannot be used for other applications. Second, the process is a highly skilled one that requires many hours of training and hence increased staffing costs. Most of the operations are also very labor intensive, particularly the collection of height data as each point has to be visited and measured individually. This contributes significantly to overall production costs.

2.6.2 The Digital Photogrammetry Era

Strongly influenced by computer technology, digital photogrammetric systems have evolved. The major difference between digital and conventional photogrammetric systems is that images used in digital systems are in digital format and hence suitable for processing by computers. If conventional aerial photographs are used, then they will need to be scanned prior to input into the system (Becker, 2001). The systems can also make use of image data collected digitally, such as satellite imagery. As with conventional analytical instruments, digital photogrammetric workstations carry out the same orientation process, discussed in the previous section, in order to model the original stereo geometry. The principles used are exactly the same, but the implementation is faster and offers greater ease of use through intuitive software interfaces.

For the time being, the biggest drawback for digital photogrammetric systems is the required analog to digital conversions (i.e., the scanning of the aerial photography). To fully transfer the information content of the aerial photography, expensive equipment and a hi-tech approach are still required. Replacing a conventional aerial survey camera with a digital equivalent becomes a challenging proposition if we take the spectral capabilities of digital cameras into consideration. Although many of the current aerial photogrammetric systems are film-based, it is expected that the use of film and conventional stereo plotters will soon be replaced by fully digital cameras and digital photogrammetric workstations. Today's digital cameras have some inherent limitations and don't produce the same resolution as film-based cameras. A standard aerial photo with

40 lp/mm corresponds to 18,400 × 18,400 pixels. Currently no CCD-chips with such a resolution are available. However, the rapid pace of digital camera evolution renders the new medium a force to be reckoned with. CCD-cameras with up to 4,000 × 4,000 pixels such as the Applanix DSS system (Mostafa, 2004), are already used in commercial applications (see Figure 2.10).

Other commercial developments include the Leica Geosystems Airborne Digital Sensor (ADS40) and the ZI Digital Mapping Camera (DMC); see Figure 2.11. The ADS40, which has been developed in cooperation with the Deutsches Zentrum für Luft-und Raumfahrt (DLR), the German Aerospace Center, is a three-line pushbroom scanner with 28° forward fore and 14° aft viewing angles from nadir. With this design, each object point is imaged three times with a stereo angle of up to 42°. Each panchromatic view direction includes two CCD lines, each with 12,000 pixels in staggered arrangement, leading to 24,000 pixels, covering a swath of 3.75 km from a flying altitude of 3,000m with a 15-cm ground pixel size. The DMC has a different design; it integrates four digital panchromatic cameras (and four multispectral 3k × 2k cameras) with CCD of 4K × 7K resolution, resulting in images with 8k × 14k resolution. With a pixel size of 12 μm × 12 μm and a focal length of 120 mm, the camera has a 43.1° × 75.4° field of view. A bundle block adjustment of a small block with crossing flight directions and with an image scale of 1:12, 800 (flying height 1,500m) resulted in a σ_0 of 2 μm (1/6 pixel). At independent checkpoints, standard deviations of $\sigma_x =$ σ_y = +/- 4 cm corresponding to 3.3 μm in the image (1/4 pixel) and σ_z = +/-10 cm corresponding to 2.7 μm x-parallax were achieved (Doerstel et al., 2002). Such accuracy is far beyond what is achievable with a film-based camera. It should be noted that although digital frame cameras do not currently reach the accuracy of film-based sensors, other digital imaging techniques surpass them by a considerable margin (Schwarz and El-Sheimy, 2004).

Figure 2.10 The Applanix DSSTM system. (Courtesy of Applanix Corporation.)

Figure 2.11 ZI Digital Mapping Camera (DMC). (Courtesy of Intergraph Corporation.)

2.6.3 Examples of Current Systems

The automatic generation of DTMs has gained much attention in the past years. A wide variety of approaches have been developed, and automatic DTM generation packages are meanwhile commercially available on several digital photogrammetric workstations (DPWS). Although the algorithms and the matching strategies used may differ from each other, the accuracy performance and the problems encountered are very similar in the major systems and the performance of commercial image matchers by far does not live up to the standards set by manual measurements (Gruen et al. 2000). One of the major difficulties in automatic feature extraction in digital photogrammetry is that of the automatic determination of image coordinates of corresponding points for the computation of object space coordinates. This topic is defined as image matching, and image matching is the location of corresponding phenomena in different images. The term of stereo matching refers to the positioning of an object in space and the reconstruction of a surface from stereo pairs (Usery, 1988; and Argialas and Harlow, 1990). The methods that are mostly used are either area-based or feature-based matching techniques using correlation of small image templates between image pairs. See Figure 2.12 for an example of a DTM generated from scanned image pairs. For more details about image matching and automated feature extraction, see Grün (1999) and Gülch (2000).

DPWS, also called softcopy workstations, have come a long way since their first appearance at the ISPRS Congress in Kyoto in 1988. Tables 2.4 and 2.5 list the characteristics of some major DPWS products. For a more complete list, see Tao (2002). Many of the current systems have a direct link between photogrammetry and GIS. Examples include the combination of Z/I's ImageStation (see Figure 2.13) with Intergraph's GeoMedia and the tight coupling between BAE Systems's SOCET SET (Figure 2.14) and ArcGIS from ESRI. The main advantage of this integration is that geometrical, topological, and semantic consistency can be achieved by combining data capture and real-time GIS analysis. In this way separate data structuring and validation steps can be avoided, and data quality is significantly increased. For more details about the state of the art of DPWS, see Heipke (2001).

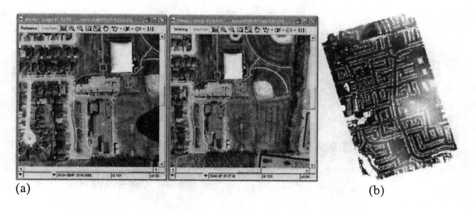

(a)   (b)

Figure 2.12 (a) A subsection of the image pairs showing automatically collected tie point on both photos, and (b) geocoded generated DTM for the whole model. (Courtesy of PCI Geomatics.)

Other trends are the direct determination of image orientation by GPS/INS and DTM generation by laser scanning (LIDAR) or interferometric radar. These developments have changed the significance of aerial triangulation and automated DTM generation by image matching and have contributed to shifting the attention more toward vector data collection and GIS integration. Due to the availability of commercial high-resolution remote-sensing imagery, integration with remote-sensing data processing becomes a clear trend in digital photogrammetry. In fact, some remote-sensing software vendors, such as ERDAS and PCI Geomatics, have been developing or integrating photogrammetric modules, and a tight integration can be expected in the near future.

Figure 2.13 Z/I ImageStation. (Courtesy of Intergraph Corporation.)

Figure 2.14 SOCET SET software. (Courtesy of BAE Systems.)

Table 2.4

Major Digital Photogrammetric Workstation (DPW) Products

Vendor	Web Site (www.)	Product Name	Number in Table 2.5
DAT/ EM Systems	datem. com	Summit Evolution	1
Leica Geo Systems (ERDAS)	erdas.com	IMAGINE OrthoBASE Pro	2
INPHO	inpho.com	Inphogrammetry	3
International System Map Corp. (I.S.M.)	askism.com	DiAP	4
Leica Geo Systems (LH Systems)	gis.leicageosystems.com	SOCET SET	5
PCI Geomatics	pcigeomatics.com	APEX 7.0	6
Z/I Imaging	ziimaging.com	ImageStation 2001	7

Table 2.5

Major Digital Photogrammetric Workstation (DPW) Products and Their Functions

Vender	1	2	3	4	5	6	7
Automatic triangulation	N	Y	Y	Y	Y	Y	Y
Automatic orientation	N	Y	Y	Y	Y	Y	Y[1]
Automatic DEM generation	Y	Y	Y	Y	Y	Y	Y
Orthoimage generation	Y	Y	Y	Y	Y	Y	Y
Mosaicking of orthoimages	Y[2]	Y[2]	Y	Y[2]	Y[2]	Y	Y
Semi-automatic feature extraction	N	Y (buildings)	N	Y	Y (lakes, buildings)	Y (linear features)	Y
Automatic contour generation	Y Fall 2002 version	Y	Y	Y	Y	Y	Y

[1] Semiautomatic interior, automatic relative orientation.

[2] Interactive and automated.

2.7 AIRBORNE LIDAR

LIDAR is an acronym for light detection and ranging, sometimes also referred to as laser altimetry or airborne laser terrain mapping (ALTM). During the last decade, laser scanning turned out to be one of the most promising techniques in photogrammetry. Some photogrammetrists even say that laser scanning has caused a paradigm change in photogrammetry because of its big impact on methods like DTM generation or 3-D city model reconstruction. A LIDAR system basically consists of integration of three technologies, namely, inertial navigation system (INS), LASER, and GPS. LIDARs are classified as active digital sensors in that they emit energy and record a returned signal, and the recorded signal is immediately converted to a digital representation and stored directly onto a computer. As active sensors, they are not dependent on sunlight and can conceivably operate 24 hours a day.

The laser ranging unit contains the laser transmitter and the receiver. The two units are mounted so that the received laser path is the same as the transmitted path. This ensures that the system will detect the target it illuminates. The target size or footprint of the laser is a function of the flying height of the platform and the divergence of the light ray. The divergence of the light thus defines the instantaneous field of view (IFOV) of the sensor. For a spatially coherent beam of laser light, the IFOV is typically between 0.3 mrad to 2 mrad (Wehr and Lohr, 1999). At a flying height of 500m, this will result in a laser footprint of 30 cm in diameter on the ground.

Solid-state lasers are now available that can produce thousands of pulses per second, each pulse having a duration of a few nanoseconds (10^{-9} seconds). The laser basically consists of an emitting diode that produces a light source at a very specific frequency. The signal is sent towards the earth where it is reflected off a feature back towards the aircraft. A receiver then captures the return pulse. Using accurate timing, the distance to the feature can be measured. By knowing a speed of the light and the time the signal takes to travel from the aircraft to the object and back to the aircraft, the distances can be computed. Using a rotating mirror inside the laser transmitter, the laser pulses can be made to sweep through an angle, tracing out a line on the ground. By reversing the direction of rotation at a selected angular interval, the laser pulses can be made to scan back and forth along a line. When such a laser ranging system is mounted in an aircraft with the scan line perpendicular to the direction of flight, it produces a saw-tooth pattern along the flight path.

The width of the strip or "swath" covered by the ranges, and the spacing between measurement points depends on the scan angle of the laser ranging system and the airplane height. Using a light twin- or single-engine aircraft, typical operating parameters are: flying speeds of 200 to 250 km per hour (55 to 70m per second), flying heights of 300 to 1,000m, scan angles generally ±20 to $\pm30°$, and pulse rates of 2,000 to 50,000 pulses per second. These parameters can be selected to yield a measurement point every few meters, with a footprint of 10 to 15 cm, providing enough information to create a DTM adequate for most applications, including the mapping of storm damage to beaches, in a single pass. The primary factor in the final DTM accuracy is the airborne GPS data. Errors in the location and orientation of the aircraft, the beam director angle, atmospheric refraction model, and several other sources degrade the coordinates of the surface point to 5 to 10 cm (Shrestha and Canter, 1998). An accuracy validation study showed that LIDAR has the vertical accuracy of 10 to 20 cm and the horizontal accuracy of approximately 1m (Murakami and Nakagawa, 1999).

Most commercial laser rangers employed in ALS systems operate in the 1,100- to 1,200-nm wavelength range (near-infrared). This is due to the common availability of lasers in that wavelength (Wehr and Lohr, 1999). The transmitted energy interacts with the target surface and permits the derivation of range and reflectance measurements. The intensity of the reflected near-infrared signal can be used to form an image of the measured area. Objects with high reflectivity such as retroreflective paint or cement contrast distinctly with objects of low reflectivity such as coal or soil. Near-infrared has the disadvantage, however, of having poor penetration of water, thus making bathymetric measurements unreliable. To achieve water penetration, a blue-green laser is used. The depth of penetration is a function of water turbidity and motion, but cannot exceed the normal penetration of blue-green light (~50m) (Optech, 2002).

To obtain a range measurement from a laser, the transmission must be modulated. There are currently two methods for modulating a laser beam for ranging: pulse modulation and sinusoidal continuous-wave (CW) modulation (Baltsavias, 1999). With pulse modulation the transmitter generates a rectangular pulse with widths from 10 to 15 ns (Wehr and Lohr, 1999). The time difference between the pulse when it leaves the transmitter and is detected in the receiver is proportional to the returned distance; that is:

$$t = 2\frac{d}{c} \tag{2.16}$$

where t is total elapsed time, d is the path distance or range of the pulse, and c is the speed of light.

Once the distances d, the laser angles α from the sensor vertical axis, the sensor attitude (from an INS system), and the sensor position (from GPS) are gathered, the position of a point on the Earth's surface can be calculated using the georeferencing equation described in Section 2.4.

2.7.1 LIDAR Scanning Techniques

There are several scanning techniques employed in different ALS systems (see Figure 2.15):

1. The first method is the use of a constant velocity-rotating mirror. This type of scanner produces measurements that appear as parallel lines on the ground. The advantage of this system is that the constant velocity does not induce any acceleration-type errors in the angle observation. The primary disadvantage is that for a certain amount of time during each mirror rotation, the mirror is not pointing at the ground and observations cannot be taken. The constant velocity mirror also restricts the FOV of the sensor, which makes the sensor less adaptable [see Figure 2.15(a)].

2. The second method is the oscillating mirror. Companies such as Optech, Leica Geosystems, and TopEye, who constitute a large portion of the commercial ALS market, employ this method. In this technique the mirror rotates back and forth. This has the effect of creating a "Z" or zigzag line of points on the ground [see Figure 2.15(b)]. The advantage of this method is that the mirror is always pointing towards the ground so data collection can be continuous. There are several disadvantages, however. The changing

velocity and acceleration of the mirror cause torsion between the mirror and the angular encoder. The changing velocity also implies that the measured points are not equally spaced on the ground. The point density increases at the edge of the scan field where the mirror slows down, and decreases at nadir.

3. The third method uses a fiber-optical array. Rather than moving a mirror to direct the laser onto the ground, a small nutating mirror is used to direct a laser into a linear fiber-optical array. The array transmits the pulse at a fixed angle onto the ground. The advantage of this system is that with fewer and smaller moving parts, the scan rate can be greatly increased. These systems typically have a sufficient scan rate such that points overlap in the along-track position. A disadvantage is that the FOV is currently much smaller than a rotating mirror and the across-track positions are fixed. Thus the only variable is the aircraft flying height.

4. The last scanning method is the Palmer or elliptical scanner. The system employs two mirrors to move the laser along an elliptical path around the aircraft. The advantage of this system is that the ground is often measured twice from different perspectives, thus allowing areas that were occluded on the first pass to be measured on the second. Disadvantages include the increased complexities of two mirrors plus the uncertainties that two angular encoders would have on a derived point location. This method was used in the ASLRIS system (Hu et al., 1999). For more details about the different operational techniques of LIDAR systems, the reader is advised to consult Morin (2003).

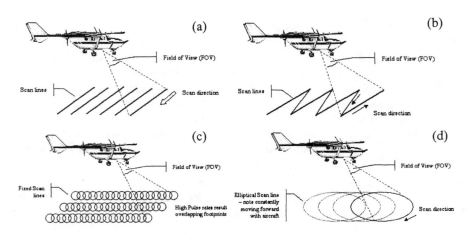

Figure 2.15 LIDAR scanning techniques: (a) constant velocity scan pattern, (b) oscillating mirror pattern, (c) nutating scanner pattern, and (d) elliptical scanner pattern. (After Morin, 2003.)

2.7.2 Special Features of LIDAR

One of the major advantages of a LIDAR system is that it can create high-accuracy DSM data. The elevation accuracy of LIDAR data is usually in the 15- to 25-cm range, making it suitable for some applications that require accurate 3-D data in urban areas such as 3-D city modeling. (See Figure 2.16 for an example of a 3-D elevation data image of the city of Toronto.) Because LIDAR systems generate 3-D coordinates of terrain points directly, the production cycle is shorter than photogrammetric methods. Table 2.6 lists the advantages and disadvantages of LIDAR data.

A feature distinguishing the principle of LIDAR from terrestrial surveying and photogrammetry, is that no selection of single prominent points takes place during measurement. LIDAR is highly cost-effective because the processing sequence of the data can be largely automated from the acquisition in flight, through the evaluation, all the way to the end product of the elevation model. The example here shows two model variants from first echo (FE) and last echo (LE), which are depicted as relief images. A digital surface model (DSM) is generated by the selection of the first echo. The last echoes are the starting point for generating the digital terrain model (DTM); see Figure 2.17.

Figure 2.16 Three-dimensional elevation data image of the city of Toronto harbor. (Courtesy of Optech Incorporated.)

Table 2.6

Advantages and Disadvantages of LIDAR Data

Advantages	Limitations
High elevation accuracy	Relatively high costs
Weather-independent to some extent	Lack of intensity image or poor-quality intensity image
Short production cycle	
Ability to partially penetrate forests	Large data volumes contain a great deal of redundancy
Ability to record multiple returns	
Can be operated at night	Inability to resolve breaklines
	Lack of standardization and commercial software

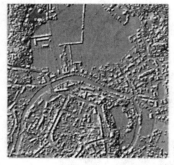   

DSM (FE)   DSM (LE)

Figure 2.17 Two model variants from first echo (FE) and last echo (LE), which are depicted as relief images. (Courtesy of TopoSys, Topographische Systemdaten GmbH.)

2.7.3 Examples of Current LIDAR Systems

A LIDAR system consists of a laser rangefinder, a GPS and IMU for determining the position and orientation of the data, a control unit, and a data storage system. LIDAR sensors emit a laser pulse, then measure the time it takes for that pulse to reflect from a terrain object and return to the sensors. The calculated distance, orientation, and attitude parameters of the system as well as the angle between the laser pulse and vertical direction determine the 3-D positions of the terrain points. Currently approximately 60 LIDAR systems are operating worldwide and several companies manufacture LIDAR equipment (Wang and Dahman, 2001). Table 2.7 lists the characteristics of a typical commercial LIDAR system while Figures 2.18 and 2.19 show the components of the TopoSys Falcon system and the Optech ALTM 3100 system, respectively. For more details about the Falcon and the ALTM 3100 system, please visit the Web pages of TopoSys GmbH (http://www.toposys.com) and Optech Incorporated (http://www.optech.on.ca), respectively.

Table 2.7

Characteristics of Typical Commercial LIDAR System

Operating range	80–3,500m nominal
Distance resolution	1–2 cm
Scan width	Variable from 0° to ±25°
Scan rate	20–70 Hz
Effective measurement rate	5,000–100,000 per sec
Laser wavelength	1500 nm
Data recording	First echo, last echo, and intensity

TopoSys Falcon LIDAR system   Falcon integrated in Piper Seneca II

Figure 2.18 The TopoSys Falcon system. (Courtesy of TopoSys, Topographische Systemdaten GmbH.)

Figure 2.19 Complete ALTM 3100 system, including control rack, sensor, laptop computer (operator interface), and pilot display. (Courtesy of Optech Incorporated.)

2.8 TERRESTRIAL LASER SCANNING SYSTEMS

The market of terrestrial laser scanners (TLS) has developed over the last few years quite successfully and the laser scanners are seen as surveying instruments that meet the requirements of detailed topographic maps and industrial applications. TLS can be considered as highly automatic motorized total stations. Unlike total stations however, where the operator directly chooses the points to be surveyed, laser scanners randomly acquire a dense set of points.

TLS provide a sampled representation of a scene by making a series of range measurements in uniform angular increments in both horizontal and vertical planes. The instrument sends a laser burst of light out of an aperture that is pointed at a known azimuth and angle above the ground plane. TLS generate clouds of data so dense that they actually begin to look like images of the objects being mapped (see Figure 2.20).

When the laser burst is reflected back, the distance from the instrument is determined and the 3-D coordinates for each measured point are derived from the basic measurements of range d, horizontal direction θ, and elevation (vertical) angle α. It is important to note that these coordinates are referenced to the instrument's internally defined system (x, y, z), which makes georeferencing into an external coordinate system (X, Y, Z) necessary.

Figure 2.20   Point cloud of Bingham Market place (Nottingham) with scanner mounted on top of a car. Five scans (different positions) merged into one cloud and grayscale derived from intensity values. (Data supplied by 3D Laser Mapping Ltd, Instrument RIEGL LMS-Z420i.)

Figure 2.21 illustrates the observables and related coordinate systems in the context of the direct scanner georeferencing. Equation (2.17) shows the georeferencing equation for calculating the coordinates of point *i* in the mapping frame.

$$r_i^m = r_o^m + R_s^m r_i^s \qquad (2.17)$$

where

$$r_i^m = \begin{bmatrix} X \\ Y \\ Z \end{bmatrix}^m \qquad (2.18)$$

are the coordinates of point *i* in the mapping frame *m*,

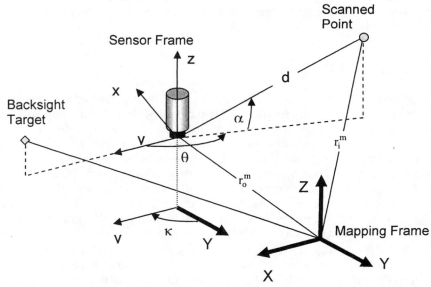

Figure 2.21 Laser scanner observables and direct georeferencing. (Adapted from Litchi and Gordon, 2004.)

$$r_i^s = \begin{bmatrix} x \\ y \\ z \end{bmatrix}^s = \begin{bmatrix} d\cos\alpha\cos\theta \\ d\cos\alpha\sin\theta \\ d\sin\alpha \end{bmatrix} \qquad (2.19)$$

are the coordinates of point *i* in the sensor coordinate system,

$$r_o^m = \begin{bmatrix} X_o \\ Y_o \\ Z_o \end{bmatrix}^m \tag{2.20}$$

are the coordinates of the TLS sensor in the mapping frame m, and

$$R_s^m = \begin{bmatrix} \cos \kappa & \sin \kappa & 0 \\ -\sin \kappa & \cos \kappa & 0 \\ 0 & 0 & 1 \end{bmatrix} \tag{2.21}$$

is the rotation matrix between the sensor frame and the mapping frame defined by the azimuth from the sensor station to the backsight station. For more details on the georeferencing and the error propagation in direct georeferencing of TLS data, see Lichti and Gordon (2004).

2.8.1 Examples of TLS Systems

At the moment, several companies are offering TLS to the market. In contrast to traditional geodetic instruments (e.g., total stations, GPS), a direct comparison of these systems is difficult because their technical specification and their physical measurement principles are different. Manufacturers of these instruments include Cyra Technologies, (a division of Leica Geosystems; see http://hds.leica-geosystems.com), Opetch (see http://www.optech.ca), and RIEGL (see http://www.riegl.com). Although their specifications vary, they all work in essentially the same manner. The systems are very portable, battery operated, require no mirrors or prisms, and use eye-safe lasers. Typical specifications include imaging distances up to 3,000 feet using powerful lasers, accuracies of around 5 to10 mm in the X, Y, Z dimensions, and field of views of up to $90° \times 360°$. Some of the TLS systems can also include a true color channel, to provide color of the target's surface as additional information to each laser measurement. An example of these systems is the RIEGL LMS-Z210i (see Figure 2.22).

Figure 2.22 RIEGL LMS-Z210i 3D Imaging Sensor. (Courtesy of RIEGL, provided by 3D Laser Mapping Ltd.)

2.9 RADAR-BASED SYSTEMS

2.9.1 Radar Imaging

RADAR (radio detection and ranging) are active remote sensing systems (as they provide their own source of illumination). Radio waves are that part of the electromagnetic spectrum that has wavelengths considerably longer than visible light. See Figure 2.23 for more details of the electromagnetic spectrum. Radar sensing has been developed on the basis of four technological principles. These are (Olmsted, 1993):

1. The ability of an antenna to emit a brief electromagnetic pulse in a precise direction;
2. The ability to detect, also with directional precision, the greatly attenuated echo scattered from a target;
3. The ability to measure the time delay between emission and detection and thus the range to the target;
4. The ability to scan with the directional beam and so examine an extended area for targets.

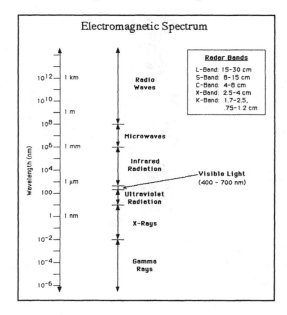

Figure 2.23 The electromagnetic spectrum.

A radar system, therefore, typically consists of a transmitter, an antenna, and a receiver. The transmitter generates a microwave pulse of short duration (effective pulse length determines range resolution). The transmitted pulse is focused into a narrow beam by an antenna to illuminate the imaged swath. The antenna receives that portion of transmitted energy signal retroreflected by the terrain/target. The receiver measures the delay and strength of the returned signal, and relates transmission/reception time delay to the range to the object from the radar. The ranging is accomplished by measuring time delay between transmission and reception of signals by knowing radio waves travel at the speed of light ($c = 3\times10^8$ m/sec). Because of the symmetry between transmission and reception patterns, the same antenna is used for both functions, with a duplex switch gating between the high-power output pulse and the low-power returned echo signal.

A detailed description of the theory of operation of radar systems is complex and beyond the scope of this section. Instead, we will provide the reader an intuitive feel for how synthetic aperture radar works. The imaging geometry of a radar system is different from other optical remote sensing systems commonly used for mapping applications (see Figure 2.24). Similar to optical systems, the platform travels forward in the flight direction (A) with nadir (B) directly beneath the platform. The microwave beam is transmitted obliquely at right angles to the direction of flight illuminating a swath (C), which is offset from nadir. Range (D) refers to the across-track dimension perpendicular to the flight direction, while azimuth (E) refers to the along-track dimension parallel to the flight direction.

This side-looking viewing geometry is typical of imaging radar systems (airborne or spaceborne). The portion of the image swath closest to the nadir track of the radar platform is called the near range (A), while the portion of the swath farthest from nadir is called the far range (B). The incidence angle is the angle between the radar beam and the ground surface (A) that increases, moving across the swath from near to far range. The look angle (B) is the angle at which the radar looks at the surface. In the near range, the viewing geometry may be referred to as being steep, relative to the far range, where the viewing geometry is shallow. At all ranges the radar antenna measures the radial line of sight distance between the radar and each target on the surface. This is the slant range distance (C). The ground range distance (D) is the true horizontal distance along the ground corresponding to each point measured in slant range. For more details, see the *ASAR User Guide* (2004).

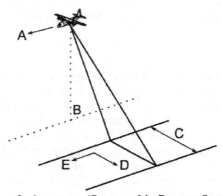

Figure 2.24 Imaging geometry of radar systems. (Courtesy of the European Space Agency.)

2.9.2 Real Versus Synthetic Aperture Radar

Both real aperture radar (RAR) and synthetic aperture radar (SAR) are side-looking systems, usually called side-looking airborne radar (SLAR), where the microwave pulse beam is radiated at an angle orthogonal to the flight direction. The difference lies in the resolution of the along-track, or azimuth direction. The length of the RAR antenna determines the resolution in the azimuth (along-track) direction of the image: the longer the antenna, the finer the resolution in this dimension.

SAR refers to a technique used to synthesize a very long antenna by combining signals (echoes) received by the radar as it moves along its flight track. SAR was invented to remove the spatial resolution dependency of RAR on the antenna length. The SAR platform configuration combined with specialized signal processing techniques synthesize an aperture that is hundreds of times longer than the actual antenna by operating on a sequence of signals recorded in the system memory. Therefore, a 1m-wide antenna can actually simulate one that is several hundred meters long. The nominal azimuth resolution for a SAR is half of the real antenna size, although larger resolutions may be selected so that other aspects of image quality may be improved. Generally, depending on the processing, resolutions achieved are of the order of 1 to 2m for airborne radars and 5 to 50m for spaceborne radars.

There are two ways to estimate terrain height with SAR; the first is to do radargrammetry based on SAR images from two passes having different viewing geometries; and the second is to compare the phases of the returns from two antennas observing the scene from (approximately) the same flight path. The second technique is called interferometry. The first technique depends on image-to-image correlation for precision and geometric accuracy of the imagery for location accuracy. Interferometry depends on estimation of the phase difference between the returns from two antennas. In principle, interferometry can be more precise, given a certain resolution, than radargrammetry.

2.9.3 IFSAR

An airborne interferometric SAR (IFSAR) system combines SAR equipment with an interferometer, two antennae, and direct georeferencing system (GPS and INS). IFSAR emits microwave signals through its two antennae and receives signals reflected from the terrain. The interferometer combines the echoes of SAR data from the two antennae to form interference patterns called an interferogram. The interferometric process has been widely discussed in the literature. Some of the general issues associated with airborne interferometry have been discussed, for example, in Gray and Farris-Manning (1993), Madsen et al. (1993), Li et al. (2002), and Mercer et al. (2003).

The basic concept of IFSAR is in having two image scenes of the same area being collected by two antennas separated in the across track (range dimension) by a small distance. The phase difference between the returns is measured. The elevations of terrain points are calculated by using the difference of phase from the same terrain object and the position of the antenna determined by onboard navigation sensors (GPS and INS). See Figure 2.25 for a schematic of the concept.

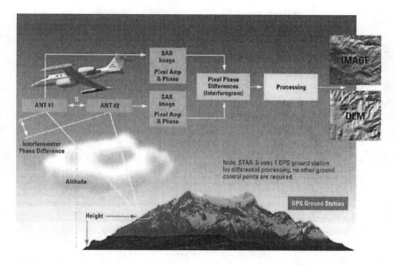

Figure 2.25 Schematic of IFSAR data processing concept. (Courtesy of Intermap Technologies Corporation.)

The geometry relevant to height extraction, *h*, is illustrated in Figure 2.26. If the two antennae, separated by baseline *B*, receive the backscattered signal from the same ground pixel, there will be a path-difference δ between the two received wavefronts. The baseline angle θ_b is obtainable from the inertial navigation system, the aircraft height is known from differential GPS, and the distance from antenna to pixel is the radar slant range. Then it is simple trigonometry to compute the target height *h* in terms of these quantities as shown in (2.22)–(2.24) (Mercer, 2004).

$$\sin(\theta_f - \theta_b) = \frac{\delta}{B} \tag{2.22}$$

$$\delta = \left[\frac{\varphi}{2\pi} + n\right]\lambda \tag{2.23}$$

$$h_i = H - r_s \cos(\theta_f) \tag{2.24}$$

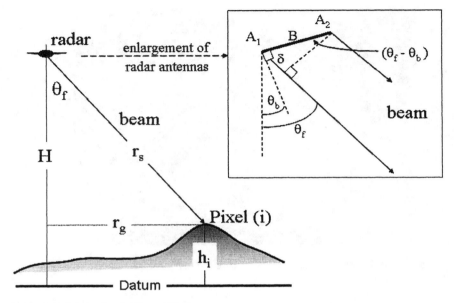

Figure 2.26 Schematic of airborne IFSAR geometry.

The path-difference δ is measured indirectly from the phase difference between the received wavefronts (2.22). The raw phase differences recorded by an interferometric radar system are known only modulus 2π (in other words, the phase differences range from 0 to 2π and then repeat). This creates an ambiguity that can only be resolved by "unwrapping" the phases by adding an integer representing the number of whole phase cycles to that point. Phase unwrapping can be a difficult process, and numerous algorithms have been proposed to perform this task (e.g., Goldstein et al., 1993). Mercer (2004), for example, suggests solving for this ambiguity with the aid of relatively coarse ground control.

2.9.4 Special Features of IFSAR

IFSAR systems are imaging systems. However, since IFSAR systems use a single frequency for illumination, there is no color associated with raw radar imagery. False-color composites of IFSAR images, however, are obtained by assigning each of the color planes in the display system to a different radar frequency and/or polarization. However, radar provides at least two significant benefits from the fact that it's not dependent on natural light, the ability to image through clouds, and the ability to image at night. The wavelength of the microwaves used in radar are longer than those of visible light, and are less responsive to the boundaries between air and the water droplets within the clouds. As well, IFSAR data can be collected at any time of day or night, and because of its wide wavelength, it can penetrate haze, clouds, water, snow, and even sand. Therefore, SAR data make a good supplement to passive image data in modern photogrammetry. The result is that, for SAR, the clouds appear homogeneous with only slight distortions occurring when the waves enter and leave the clouds.

The images they generate are useful for monitoring shoreline erosion, investigating ancient rivers beneath desert sands, studying glaciers, and mapping snow-covered rock formations. SAR's long wavelength is more sensitive to the physical properties, shape, and size of a sensed object than it is to color and chemical composition (Wang and Dahman, 2001).

2.9.5 Examples of Current IFSAR Systems

Currently several airborne IFSAR systems are in operation around the world. IFSAR elevation accuracy is 1 to 5m and the spatial posting of the data usually ranges from 2.5 to 10m. Many IFSAR vendors develop their own software to automatically generate bald-earth DTMs from the original DSM by removing ground objects such as buildings and sparse trees.

Two examples of IFSAR implementation are STAR-3i and TopoSAR, which are both deployed operationally by Intermap Technologies Corporation, Canada. A STAR-3i was originally designed and built by ERIM, but has subsequently had major upgrades in hardware and software (Tennant et al., 2003); see Figure 2.27. TopoSAR was originally developed by AeroSensing under the name AeS-1 (Hofmann et al., 1999) and has also experienced upgrades—mostly in the processing area. See Table 2.7 for the characteristics of the two systems (Mercer, 2004).

Figure 2.27 The Intermap's STAR-3i system. (Courtesy of Intermap Technologies Corporation.)

Table 2.7

Typical Operating Parameters of Intermap's STAR-3i System

Typical Parameters	STAR-3i	TopoSAR	
Platform	Lear Jet	AeroCommander	
Altitude (km)	6.5–9.5	3.5–6.5	
Speed (km/h)	700	450	
Frequency band	X	X	P
Center wavelength (cm)	3	3	74
Image resolution (m)	1.25	0.5 – 2	2
Polarization	HH	HH	HH, W, HV/VH
Swath width (km)	5, 10	2, 4, 7	4
IFSAR mode	Single-pass	Single	Repeat-pass
DEM spacing (m)	5	1, 2.5, 5	2.5

Source: Mercer, 2004.

The specifications of the STAR-3i derived DTM and orthorectified radar image (ORI) are summarized in Table 2.8. Varying flying altitudes and operating modes enable different accuracy specifications to be achieved that may be reflected in cost and other factors. For more details, please visit the Intermap Technologies Corp. Web site (http://www.intermaptechnologies.com). Figure 2.28 shows an image of Intermap's NEXTMap Britain mapping program. This image is a mosaic of over 2,800, 10 × 10-km tiles of the digital surface model, which were resampled to 50m postings and shaded to highlight relief.

Table 2.8

Intermap Core Product Specifications for STAR-3i DSM and DTM
(http://www.intermaptechnologies.com)

Product Type	DSM (m)		DTM (m)	
	RMSE[1]	Spacing	RMSE	Spacing
I	0.5	5	0.5	5
II	1	5	1	5
III	3	10	—	—

[1] RMSE refers to vertical accuracy.

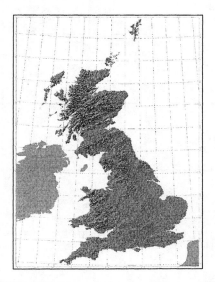

Figure 2.28 Intermap's NEXTMap Britain mapping program. (Courtesy of Intermap Technologies Corp.)

2.10   COMPARISON OF DIFFERENT METHODS

There is no question that current methods for generating DTM data meet the accuracy and resolution requirements for all the applications described in Chapter 1. However, the question has always been about how to produce low-cost, high-accuracy, high-resolution DTMs. The cost for generating DTMs using conventional approaches can become significant for increased resolution, accuracy, and especially number of elevation postings. Costs associated with various methods of generating DTMs are directly reflective of the labor-intensive and time-consuming processes currently used. Figure 2.29 shows a comparison of the cost (in U.S. dollars) of producing 1 km^2 against the accuracy of the different techniques presented in this chapter (for the sake of completeness, other techniques such as satellite imagery are also included in the figure). Other factors that decide on the choice of the technique to use include the other data that will be needed with the DTM, such as images. This might dictate the choice of the technique.

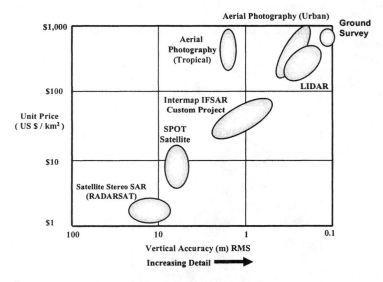

Figure 2.29 Unit cost comparison of DTMs as a function of typical vertical accuracies addressed by various technologies. (Adopted from Mercer, 2004.)

References

Argialas, D.P., and Harlow, C.A., "Computational image interpretation models: an overview and a perspective," *Photogrammetric Engineering and Remote Sensing*, 56(6): 871-886, 1990.

ASAR User Guide, The European Space Agency (ESA), http://envisat.esa.int/dataproducts/asar/toc.htm: 2004.

Baltsavias, E.P., "Airborne laser scanning: basic relations and formulas," *ISPRS Journal of Photogrammetry & Remote Sensing*, 54: 199-214, 1999.

Bäumker, M. and A. Mattissek, "Integration of a Fibre Optical Gyro Attitude and Heading Reference System with Differential GPS," *Proc. ION GPS-92*, Albuquerque, NM, USA, Sept. 16-18, 1992, pp. 1093 – 1101.

Becker, R., "The digitalization of photogrammetry investing in, and exploiting data," *Digital India, Hyderabad*, July 6-7, 2001.

Benson, D.O., Jr., "A comparison of two approaches to pure-inertial and doppler-inertial error analysis," *IEEE Transactions on Aerospace and Electronic Systems*, AES-11 (4): 1975.

Cannon, M.E., and Schwarz, K.P., "A discussion of GPS/INS integration for photogrammetric applications," *Proc. IAG Symp. # 107: Kinematic Systems in Geodesy, Surveying and Remote Sensing*, 443-452, New York: Springer Verlag, Sept. 10-13, 1990.

Chavez, P.S., Jr., "Atmospheric, solar, and MTF corrections for ERTS digital imagery," *Proceedings of the American Society of Photogrammetry, Fall Technical Meeting*, Phoenix, AZ, 69-69a, Oct. 1975.

Cosandier, D., Chapman, M.A., and Ivanco, T., "Low cost attitude systems for airborne remote sensing and photogrammetry," *Proc. of GIS93 Conference*, Ottawa: 295-303, March 1993.

Crassidis, J.J., and Markley, F.L., "Unscented filtering for spacecraft attitude estimation," *J. of Guidance, Control, and Dynamics*, 26(4): 536-542, 2003.

Doerstel, C., Zeitler, W., and Jacobsen, K., "Geometric calibration of the DMC: method and results," *ISPRS Com I and PECORA*, Denver, CO, 2002.

El-Sheimy, N., and Schwarz K.P., "Kinematic positioning in three dimensions using CCD technology," *Proc. IEEE/IEE Vehicle Navigation & Information System Conference (IVHS)*, 472-475, Oct. 12-15, 1993.

El-Sheimy, N., *The Development of VISAT — A Mobile Survey System for GIS Applications*, Ph.D. thesis, UCGE Report No. 20101, Department of Geomatics Engineering, the University of Calgary, 1996.

Glasbey, C.A., "An analysis of histogram-based thresholding algorithm," *Graphical Models and Image Processing*, 55(6): 532-537, 1993.

Goldstein, R.M., H.A. Zebker, and C. Werner, "Satellite radar interferometry: two dimensional phase unwrapping," *Radio Sci.*, 23(4): 713-720, 1993.

Gonzalez, R.C., and Wintz, P., *Digital Image Processing*, Reading, MA: Addison-Wesley, 1987.

Gray, L.A., and Farris-Manning, P.J., "Repeat-pass interferometry with airborne synthetic aperture radar," *IEEE Transactions on Geoscience and Remote Sensing*, 31(1): 180-191, 1993.

Gruen, A., Bär, S., and Bührer, Th., "DTMs derived automatically from DIPS — where do we stand?" *Geoinformatics*, 3(5): 36-39, 2000.

Grün, A., "Digital photogrammetric stations," *GeoInformatics*, 2: 22-27, 1999.

Gülch, E., "Digital systems for automated cartographic feature extraction," *IAPRS* Vol. XXXIII, Part B2: 241-256, 2000.

Heipke, C., "Digital photogrammetric workstations — a review of the state-of-the-art for topographic applications," *GIM International*, 15(4): 35-37, 2001.

Heipke, C., Englisch, A., Speer, T., Stier, S., and Kutka, R., "Semi-automatic extraction of roads from aerial images," *Proceedings of International Society of Photogrammetry and Remote Sensing*, Commission III Symposium, Munich: 353-360, 1994.

Hofmann, C., Schwabisch, M., Och, S., Wimmer, C., and Moreira, J., "Multipath p-band interferometry – first results," Proceedings of the Fourth International Airborne Remote Sensing Conference and Exhibition, 21st Canadian Symposium on Remote Sensing, Ottawa, Ontario, Canada, 1999.

Hu, Y., Xue, Y., Fang, K., and Pan, Z., "Scanning laser altimeter in airborne scanning laser ranging-imaging *sensor," Proceedings of the Fourth International Airborne Remote Sensing Conference and Exhibition,* Canadian Symposium on Remote Sensing, Ottawa, Canada I, 510-517, 1999.

Jawahr, C., Biswas, P., and Ray, A., "Analysis of fuzzy thresholding schemes," *Pattern Recognition,* 33(8): 1339-1349, 2000.

Julier, S.J., and Uhlmann, J.K., *A General Approach for Approximating Nonlinear Transformations of Probability Distributions,* Technical Report, Department of Engineering Science, University of Oxford, United Kingdom: 1996.

Julier, S.J., and Uhlmann, J.K., "The scaled unscented transformation," *Proc. IEEE American Control Conference,* Anchorage, AK: 4555-4559, 2002.

Kreifelts, T. "Skelettierung und Linienverfolgung in rasterdigitalisierten Linienstrukturen," in H.H. Nagel, Hg., Digitale Bildverarbeitung, Bd. 8 von Informatik Fachberichte, S. 223-231, New York: Springer, 1977.

Lapucha, D., Schwarz, K.P., Cannon, M.E., and Martell, H., "The use of GPS/INS in a kinematic survey system," *Proc. IEEE PLANS* 1990, Las Vegas, NV, March 20-23, 1990, pp. 413-420.

Leberl, F., and Olson, D., "Raster scanning for operational digitizing of graphical data," *Photogrammetric Engineering and Remote Sensing,* 48(4): 615-627, 1982.

Lemmens, M., "A survey on stereo matching techniques," *International Archives of Photogrammetry and Remote Sensing,* Commission V, 27(B8): 11, 1988.

Li, X., Baker, B., and Hutt, T., "Accuracy of airborne IFSAR mapping," *Proceedings of the XXII FIG International Congress,* ACSM/ASPRS Annual Conference, Washington, D.C., April 9-26, 2002.

Lichti, D., and Gordon, S., "Error propagation in directly georeferenced terrestrial laser scanner point clouds for cultural heritage recording," *FIG Working Week 2004,* Athens, Greece, May 22-27, 2004.

Lipman J.S., "Trade-Offs in the Implementation of Integrated GPS Inertial Systems," *Proc. ION GPS-92,* Albuquerque, NM, Sept. 16-18, 1992, pp. 1125-1133.

Madsen, S., Zebker, H.A., and Martin J., "Topographic mapping using radar interferometry: Processing techniques," *IEEE Transactions on Geoscience and Remote Sensing,* 31: 246-256, 1993.

Mercer, B., Allen, J., Glass, N., Reutebuch, S., Carson, W., and Andersen, H.E., "Extraction of ground DEMs beneath forest canopy using fully polarimetric InSAR," *Proceedings of the ISPRS WG I/3 and WG II/2 Joint Workshop on Three-Dimensional Mapping from InSAR and LIDAR,* Portland, OR, (CD) 2003.

Mercer, B., "DEMs created from airborne IFSAR — an update," *The International Society for Photogrammetry and Remote Sensing (ISPRS) Congress,* Commission II, WGII/2, Istanbul, Turkey: (7 Pages, CD), July 15-22, 2004.

Morin, K.W., "Calibration of Airborne Laser Scanners," M.Sc. Thesis, Department of Geomatics Engineering, the University of Calgary, Calgary, Canada, 2003.

Mostafa, M.M.R., and Schwarz, K.P., "A multi-sensor system for airborne image capture and georeferencing," *PE&RS,* 66(12): 1417-1424, 2000.

Mostafa, M., Hutton, J., and Lithopoulos, E., "Ground accuracy from directly georeferenced imagery, *GIM International,* 14(12): 2000.

Mostafa, M., "The digital sensor system data flow," *The 4th International Symposium on Mobile Mapping Technology*, Kunming, China (CD), April 2004.

Murakami, H., and Nakagawa, K., "Change detection of buildings using an airborne laser scanner," *ISPRS Journal of Photogrammetry and Remote Sensing*, 54: 148-152, 1999.

Olmsted, C., Scientific SAR User's Guide, Alaska SAR Facility, http://www.asf.alaska.edu/SciSARuserGuide.pdf, 1993.

Optech Inc., *Optech – Laser Based Ranging, Mapping and Detection Systems*, Optech Incorporated, http://www.optech.on.ca, 2002.

Pal, S.K., "A review on image segmentation techniques," *Pattern Recognition*, 26(9): 1277-94, 1993.

Scherzinger, B.M., "Inertial navigator error models for large heading uncertainty," *Proc. Int. Symp. Kinematic Systems in Geodesy, Geomatics and Navigation*, Banff, Canada, 121-130, Aug. 30-Sept. 2, 1994.

Schwarz, K.P., "Mapping the earth's surface and its gravity field by integrated kinematic systems," Lecture Notes of the Nordic Autumn School, Fevic, Norway: Aug. 28-Sept. 1, 2000.

Schwarz, K.P., Chapman, M.A., Cannon, M.E., and Gong, P., "An integrated INS/GPS approach to the georeferencing of remotely sensed data," *PE&RS* 59 (11): 1667-1674, Nov. 1993.

Schwarz, K.P., and El-Sheimy, N., "Mobile mapping technologies: state of the art and future trends," *The International Society for Photogrammetry and Remote Sensing (ISPRS) 2004 Congress*, Commission I, Istanbul, Turkey, July 15-22, 2004.

Shears, J.C., and Allan, J.W., "Softcopy photogrammetry and its uses in GIS," *ESRI User Conference*, Palm Springs, CA, May 20-24, 1996.

Shrestha, R., and Canter, W.E., "Engineering applications of airborne scanning lasers: reports from the field," *Journal of Photogrammetry Engineering and Remote Sensing*, 66: 256, 1998.

Shin, E.H. and El-Sheimy, N., "An unscented Kalman filter for in-motion alignment of low-cost IMUs," *IEEE PLANS*, Monterey, CA, 273-279, April 26-29, 2004.

Shortridge, A.M., "Characterizing uncertainty in digital elevation models," Chapter 11 in *Spatial Uncertainty in Ecology: Implications for Remote Sensing and GIS Applications*, C.T. Hunsaker, M.F. Goodchild, M.A Friedl, and T.J. Case, (eds.), New York, Springer, 238-257, 2001.

Tao, V., "Digital photogrammetry: the future of spatial data collection," (*Technology Trends*), *GEO World*: 30-37, May 1, 2002.

Tennant, J.K., Coyne, T., and DeCol, E., "STAR-3i interferometric SAR (InSAR): more lessons learned on the road to commercialization," *Proceedings of the ASPRS Conference*, Charleston, SC, (CD-ROM), 2003.

Usery, E.L., and Altheide, P., "Knowledge-based GIS techniques applied to geological engineering," *Photogrammetric Engineering and Remote Sensing*, 54(11): 1623-1628, 1988.

Wang, Y., and Dahman, N., "Active sensors and modern photogrammetry," *Geospatial Solutions*, 2001, www.geospatial-online.com/geospatialsolutions/article/articleDetail.jsp?id=10243.

Wehr, A., and Lohr, U., "Airborne laser scanning: an introduction and overview," *ISPRS Journal of Photogrammetry & Remote Sensing*, 54: 68-82, 1999.

Weszka, J.S., "Survey of threshold selection techniques," *Computer Graphics and Image Processing*, 7(2): 259-265, 1978.

Wu, Y., "R2V conversion: why and how?" *GeoInformatics*, 6(3), 28-31, 2000.

Xin, Y., *Automating Procedures on a Photogrammetric Softcopy System*, M.Sc. Thesis, Department of Geomatics Engineering, the University of Calgary, Calgary, Canada, 1995.

Chapter 3

DTM Data Structures

3.1 INTRODUCTION

A DTM may take the form of irregularly spaced points, contours, a grid, or a TIN (see Chapter 1). Irregularly spaced points are a very common format for initial observations of elevation, (for example, in image matching from stereo images), but cannot truly be considered a model of the surface. This is because it does not contain information (either explicit or implicit) about the nature of the surface between the data points. The remaining three methods are, in principle, commonly derived from irregular point data. A TIN, for example, connects the points together to form a set of triangular facets covering the entire surface, and it is these triangles that define the nature of the surface and allow for the estimation of elevation and surface derivatives at any point. Contours and grids involve interpolation from the original data; in the case of contours, by keeping the elevation constant and interpolating the location, and in the grid data case, by defining the location and interpolating the elevation (see Chapter 4 for the different interpolation techniques). Furthermore, contour and grid representations can be collected directly from photogrammetric stereo models as has been discussed in Chapter 2. In this chapter we will discuss in more detail the data structure of grids and TINs, as they are the most commonly used data format for storing DTM data.

3.2 DATA STRUCTURE IN GRID DATA

Regular grid models have been the most used representation for terrain surfaces. This is because grid models are based on a very simple data structure. The regular grid data structure is comprised of a 2-D matrix of elevations, sampled at regular intervals in both the x and y planes. The term regular grid suggests that there is a formal topological structure to the elevation data. In reality, the grid data structure is only representative of discrete points in the *x-y* planes (Figure 3.1) and should not be thought of as a continuous surface. However, it is often perceived as such when combined with methods of surface interpolation such as those presented in Chapter 4. Generally speaking, the grid surface is a model that fits the data structure. On the other hand, the TIN data structure is made to fit the surface model. The grid model is a set of points connected by vectors of fixed length and direction, while, in the triangular model, the length and direction of the connection vectors are both variable. Because of the fixed vectors, the height values in the grid model can be stored in a matrix. In some special grid models, the length of the connection vector may be partly variable.

The spatial resolution of a grid DTM is inherently constrained to cell size — the smaller the cell size, the higher the resolution, which is apart from the quality of the original data. However, once generated, the source data is lost and no further improvement is possible: one can only downsample a grid DTM (i.e., go to a larger cell size and, thus, lower resolution). Creating a new grid of a smaller cell size out of an existing grid DTM will not increase its spatial resolution. A TIN DTM, on the other hand, does not suffer from this constraint due to its adaptive nature, although a small elevation tolerance may be employed to reduce the amount of data used in constructing a TIN. Many mapping agencies around the world, for example the United States Geological Survey (USGS), choose grid representation for their terrain data due to its simplicity and relatively small storage size. The most common source of digital terrain elevation data is the gridded digital elevation model (DEM). The USGS produces and distributes DEMs covering the entire United States at a variety of scales. These DEMs are derived from contour maps, stereo-models of high-altitude photographs, or the digital line graph (DLG) hypsography data, another digital source that includes contours (USGS, 1990). The Defense Mapping Agency (DMA) produces DEMs with worldwide coverage at a variety of scales as well, although distribution is limited. Some of the digital DMA products in this category are digital terrain elevation data (DTED), world mean elevation data (WMED), and digital bathymetric data base (DBDB) (DMA, 1990). DTED is the most commonly used of these. It forms the basis of many other DMA digital products.

Some mapping/DTM software store grid data in a hierarchical tile-block structure. In this structure, data is stored by a combined hierarchical and relational data model. The grid is first divided into uniform square units called tiles. Each tile represents an actual portion of geographic space. A tile is further divided into blocks. A block is made up of cells arranged in a grid form; see Figure 3.1. For more details about hierarchical tile-block structure, see Peng et al. (2004).

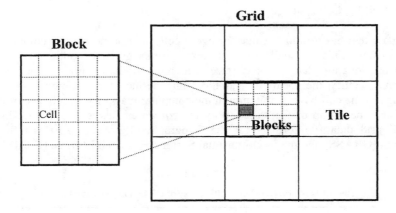

Figure 3.1 Grid tile-block data structure.

The tile-block structure allows random access to data and consequently rapid retrieval of information maintained from any subsection of a grid, regardless of the size of the database. The tile structure is for spatial indexing; it does not divide data into discontinuous map sheets. The data is accessed through a hierarchy of file–tile–block–cell pointers, allowing for random access and rapid retrieval of information, regardless of the size of the database.

3.3 STORAGE AND COMPRESSION TECHNIQUES OF GRID DATA

Since DTM (especially grid) files are so large, minimizing storage space is important. Although storage space continues to become cheaper, the data continues to have higher and higher resolutions. Besides the cost of the space, the data file size also affects transmission time. Data compression techniques can be applied to reduce the size of grid databases. However, at the same time this means a loss of information if a "lossy" compression technique is used. Data consisting of large homogeneous features can be compressed better than data consisting of small fragmented polygons. Specific compression techniques compress some data better than others, but no compression technique compresses all types of data more effectively than another.

In this section we will briefly introduce the basics of some of the main compression techniques. Only the theory behind the compression technique is discussed; the implementation of the technique may vary greatly from system to system. The techniques that will be addressed are "cell by cell" (uncompressed), "run-length code," and "quad trees."

3.3.1 Cell-by-Cell Storage

For data that has a unique value for every cell, there is no way to compress the information. This is usually true for floating-point grids or continuous surfaces such as elevation data, slope surface, and noise pollution grids. The row and column identify the location, and the value defines the attribute. The actual storage of the cell-by-cell structure in the computer may vary widely according to the implementation of the structure by the software. Tables 3.1 and 3.2 list the DTM grid data file formats for the Geographic Resources Analysis Support System (GRASS) and the Environmental Systems Research Institute (ESRI) Inc.

Table 3.1

The GRASS Grid Data Format (Courtesy of GRASS Development Team)

Northern boundary grid coordinate	north: n
Southern boundary grid coordinate	south: s
Eastern boundary grid coordinate	east: e
Western boundary grid coordinate	west: w
Number of rows in the grid	rows: nrows
Number of columns in the grid	cols: ncols
Values of row 1	$z_{11} \; z_{12} \; z_{13} \; ... \; z_{1ncols}$
Values of row 2	$z_{21} \; z_{22} \; z_{23} \; ... \; z_{2ncols}$
……	…….
……	…….
Values of last row	$z_{nrows1} \; z_{nrows2} \; z_{nrows3} \; ... \; z_{nrowsncols}$

Table 3.2

ESRI DTM Grid Data Format (Courtesy of ESRI Inc.)

ncols ncol	Number of columns in the grid
nrows nrow	Number of rows in the grid
xllcorner x	Lower left x coordinate of grid
yllcorner y	Lower left y coordinate of grid
cellsize size	Grid cell size
NODATRA_value NODATA	Value of an empty grid cell
$z_{11} z_{12} z_{13} \dots z_{1ncols}$	Values of row 1
$z_{21} z_{22} z_{23} \dots z_{2ncols}$	Values of row 2
.	
.	
$z_{nrows1} z_{nrows2} z_{nrows3} \dots z_{nrowsncols}$	Values of last row

3.3.2 Run-Length Code

In run-length encoding, adjacent cells along a row that have the same value are treated as a group termed a run. Instead of repeatedly storing the same value for each cell, the value is stored once, together with information about the size and location of the run. In standard run-length encoding, the value of the attribute, the number of cells in the run, and the row number are recorded. It is a "lossless" compression, as opposed to a "lossy" compression since the original data can be exactly reproduced.

Run-length coding stores data by row. A row is identified first, followed by the columns and the value associated with the row-column location. The column range information is stored as a from-column to a to-column along with the value that is associated with that cell or group of cells. If two or more adjacent cells have the same value, only the from- and to-columns and the value are stored. The more columns that can be included in the from- and to-column identification as having the same value, the greater the compression will be. In standard run-length encoding, the value of the elevation, the number of cells in the run, and the row number are recorded. Another type of run-length encoding data compression is the value point encoding, in which cells are assigned to position number starting in the upper left corner, proceeding from left to right and from top to bottom. The position number for the end of each run is stored in the point column. The value for each cell in the run is in the value column. Table 3.3 shows an example of how to represent grid data using a run-length code.

While run-length coding is very simple to implement, it has a major drawback in terms of data access time. As opposed to the standard way of accessing grid data (by simply counting the number of cells across a row to locate a given cell), access to a cell in run-length coding must be computed by summing along the run-length codes. This is typically a minor additional cost, but in some applications the trade-off between speed and data volume may be objectionable.

Table 3.3

An Example of Grid Data Representation Using Run-Length Coding

DTM Data										Run-Length Coding
1	2	3	4	5	6	7	8	10	11	
101	101	101	101	102	102	102	102	103	103	101, 4, 102, 8, 103, 11
101	101	101	101	103	104	104	104	104	104	101, 4, 103, 5, 104, 11
101	101	101	101	104	104	104	104	104	105	101, 4, 104, 10, 105, 11
102	102	102	103	105	105	105	105	106	106	102, 3, 103, 4, 105, 8, 106, 11

3.3.3 Quad Tree

The quad tree data model provides a more compact raster representation by using a variable-sized grid cell. Instead of dividing an area into cells of one size, finer subdivisions are used in those areas with finer detail. By doing this, a higher level of resolution is provided only where it is needed. The process typically starts by dividing a grid area into four equal, smaller squares when cells in the grid have different values. If all the cells within any of the four smaller squares have the same value, the square will not be broken down any further. However, if there are cells with different values (therefore the square does not consist of all homogenous cells), that square has to be subdivided into four more equal squares. This process continues until each square only represents cells with the same values.

The quad tree is so named because the squares are stored in a hierarchical tree. The tree starts with one branch and then produces four branches (or four children). Each time a square is divided, it is always divided into four more branches (children), and the quad tree grows a level. When the quads or squares no longer need to be subdivided, a value is assigned to represent each branch and level. There are many ways to structure the data pointers, and these methods are beyond the scope of this section; however, for further information on the structure of quad trees the reader is advised to consult Samet (1984) and Chen and Tobler

(1986). Figure 3.2 shows an example of the quad tree representation of an area of size X. Each node is represented by at least four grid elevations, which correspond to squares. The root node represents the whole terrain data with four corners. As we go down to the deeper levels in the quad tree hierarchy, the distance between the corners is halved and at the deepest level there remains no other elevation point that is to be represented by the nodes of the quad tree (Figure 3.2). The data structure obtained after the quad tree scheme is usually passed to a simplification algorithm to eliminate the nodes having the same elevations within all children.

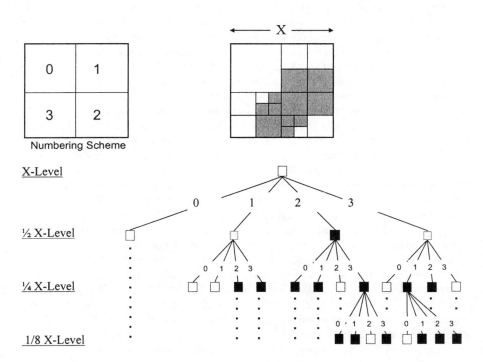

Figure 3.2 Quad tree structure.

Quadrants are often numbered from 0 to 3. There is no standard numbering scheme, however. The physical structure of the computer file is also organized according to the chosen numbering scheme. As a result, cells that are close together on the grid are stored close together in the file as well. For operations that require data for a neighborhood, this storage organization provides efficient data retrieval. Quad trees are particularly efficient for identifying the nearest neighbor of a selected point and for identifying the area (polygon) in which a point is located (point-in-polygon search). The major disadvantage of quad trees is the time it takes to create and modify them. Quad trees can provide efficient storage of data but only if the data is fairly homogenous. The fewer the classes and the larger the cluster, the greater the degree of compression and the more efficient the quad tree structure.

3.4 DATA STRUCTURES FOR TINS

Unlike the grid data, which is readily available from different vendors, terrain data is not readily available in the form of a TIN. The TIN model has to be generated from either irregularly distributed points or grid data. In the TIN model, information about the connection vector is implicit in the data structure. A triangle is formed by three edges where each edge is defined by two terrain elevation endpoints. It is important to notice that an endpoint might be shared among two or more triangles while an edge might be shared among one or two different triangles. Various combinations of the connection between these three types of elements (nodes, edges, and triangles) form the TIN data structure.

Furthermore, unlike grid data in which each point (or node) is connected to either four or eight adjacent neighbors, the number of neighbors connected to a given node in a TIN may in theory be arbitrarily large. Ideally, a data structure should represent this variable connectivity in a way that (1) provides rapid access to adjacent mesh elements without demanding excessive storage space, and (2) is flexible enough to handle dynamic changes in the mesh itself. For dynamic modeling applications, an additional requirement is the need to maximize computational speed. There are several models that satisfy these requirements; see, for example, the "dual edge" or the "quad edge" data structure of Guibas and Stolfi (1985) and the twin-edge (dual-edge) structure proposed by Heller (1990). Both models consist of three geometric elements: nodes, triangles, and edges [or directed edges as called in the Guibas and Stolfi (1985) model]. It is important to be aware of the triangulation algorithm (the technique to construct the TIN model) that is going to be used when the data structure is chosen (see Section 3.8 for the different triangulation algorithms). It is also useful to know what type of information the surface model is to include, and what the model is intended for. Table 3.4 lists the geometrical and topological information in a TIN data structure while the next three sections will discuss these structures in more detail.

Table 3.4

The Geometrical and Topological Information in a TIN Data Structure

TIN Data Structure	Geometrical Information	Topological Information
Nodes	x, y, and z coordinates	Pointers to adjacent edges Pointers to adjacent triangles Pointers to surrounding nodes
Edges	Coordinates of the endpoints/ pointers to the endpoints	Pointers to adjacent edges Pointers to adjacent triangles
Triangles	Coordinates of the vertices/ pointers to the vertices of the triangle	Pointers to the edges of the triangle Pointers to adjacent triangles

3.5 NODE-BASED DATA STRUCTURE

Data sets for terrain surfaces very often consist of large amounts of points. Consequently it is important to use an efficient node handling structure during construction of the TIN. The running time of the triangulation algorithms actually depend to a large degree on the point handling structure. There are several ways through which the node structure can be presented. The following is one of them — each node data object includes (x, y, z) coordinates, the number of neighboring nodes, and a pointer to one of its directed edges (that is, one of the directed edges that originates at the node, here referred to as a "spoke" of that node). Note that a pointer to a single spoke is all that is needed to fully describe the connectivity among nodes. Because each spoke points to its counterclockwise neighbor, a list of spokes and neighboring nodes can be easily constructed for any node. For finite-difference applications, a node object can also include geometric information such as the projected surface area of its Voronoi diagram (Voronoi area) and a flag indicating whether it lies on the boundary or interior of the mesh (Tucker and Slingerland, 1997).

3.6 TRIANGLE-BASED DATA STRUCTURE

A triangle-based data structure is based on three tables (as described in Bjørke, 1988): a point table containing $x, y,$ and z coordinates; an edge table with pointers to the endpoints and adjacent triangles of the edges; and a triangle table consisting of pointers to the edges that define the current triangle. Figure 3.3 shows a simple triangular mesh. Indices of the edges, points, and triangles are indicated in the mesh. The tables show the data structure for the current mesh. In this data structure the neighborship between edges is represented by the triangle table.

Gaustad (1990) made use of a triangle-based structure for triangulation by an incremental method. For each triangle he stored pointers to the vertices of the current triangle (always 3), and pointers to adjacent triangles (0 – 3). There is also a pointer to the list of points that lie inside each triangle. Depending on the application, additional geometric data for triangles might include projected area, gradient, and gradient vector (see, for example, Palacios-Velez and Cuevas-Renaud, 1986).

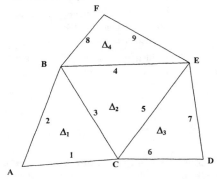

Edge	1st Δ	2nd Δ	1st point	2nd point
1	Δ1	0	A	C
2	Δ1	0	A	B
3	Δ1	Δ2	B	C
4	Δ2	Δ4	B	E
5	Δ2	Δ3	C	E
6	Δ3	0	C	D
7	Δ3	0	D	E
8	Δ4	0	B	F
9	Δ4	0	E	F

Δ No.	1st edge	2nd edge	3rd edge
Δ1	1	2	3
Δ2	3	4	5
Δ3	5	6	7
Δ4	4	8	9

Figure 3.3 Triangle-based data structure.

3.7 EDGE-BASED DATA STRUCTURE

In an edge-based data structure all information about the triangle is implicit in the edge table. Consequently, no triangle table is required. One of the edge-based data structures was proposed by Heller (1990) and is called the twin-edge (dual-edge) structure. The neighborship in this model is given only by the edge table. A pointer to the endpoint and a pointer to the next edge in the triangle are stored for each edge. Further a pointer to the twin-edge is stored. The twin edge is geometrically identical to the current edge, but it points at the opposite endpoint. In addition, there is a pointer from each edge to the point list belonging to the triangle. In the twin-edge structure a single "edge record" is a part of only one triangle. A triangle is formed by three separate edges. To make the twin-edge structure comparable to the triangle-based structure, the triangular mesh from Figure 3.3 is shown for the twin-edge structure in Figure 3.4.

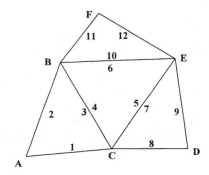

Point No.	X Y Z	Next
A	.. .. ..	
B	.. .. ..	
C	.. .. ..	
D	.. .. ..	
E	.. .. ..	
F	.. .. ..	

Edge	Point	Next	Twin	Attribute
1	A	2	0	....
2	B	3	0	....
3	C	1	4	....
4	B	6	3	....
5	C	4	7	....
6	E	5	10	....
7	E	9	5	....
8	C	7	0	....
9	D	8	0	....
10	B	11	6	....
11	F	12	0	....
12	E	10	0	....

Figure 3.4 Edge-based data structure.

3.7.1 Voronoi Diagrams and Delaunay Triangulation

A Voronoi diagram (VD) is a geometric structure that represents proximity information for a set of points or objects. Given a set of sites or objects, a VD represents the regions of a plane that are closer to a particular point in the plane than to any other point. That is, the points on the Voronoi diagram are equidistant to two or more sites (see Figure 3.5). The VD is named after the mathematician M. G. Voronoi who explored this geometric construction in 1908 (Voronoi, 1908). Voronoi diagrams were first discussed by Dirichlet (1850). Accordingly, the VDs are sometimes also called Thiessen polytopes or Dirichlet regions.

Voronoi diagrams are of major importance in computational geometry problems. For example, Voronoi diagrams can be used to solve the nearest neighbor problem: given a set of data points representing samples of objects from several classes and the point associated with an object whose type is unknown, we wish to determine into which class the unknown object falls. If we delineate the Voronoi regions around each existing point then the unknown's type can be identified as the same as the object responsible for the Voronoi region in which the unknown's data point lies.

Delaunay triangulation is closely related to the Voronoi diagram. The triangulation is named after B. Delaunay, who first made use of the dual relationship (Delaunay, 1934). If we use the Voronoi diagram as a basis, we can construct the Delaunay triangulation by drawing the lines between the points in adjacent polygons. When the construction is finished we have a triangular network that covers the whole area. The relationship between the Voronoi diagram and the Delaunay network is shown in Figure 3.5.

The Delaunay triangulation of a point set is a collection of edges satisfying an "empty circle" property: for each edge we can find a circle containing the edge's endpoints but not containing any other points. The Delaunay triangulation is the dual structure of the Voronoi diagram in R^2. By dual, we mean to draw a line segment between two Voronoi vertices if their Voronoi polygons have a common edge, or in mathematical terminology, there is a natural bisection between the two that reverses the face inclusions. The circumcircle of a Delaunay triangle is called a Delaunay circle. Delaunay triangulations have a number of interesting properties that are consequences of the structure of the Voronoi diagram:

1. Convex hull: The exterior face of the Delaunay triangulation is the convex hull of the point set.
2. Circumcircle property: The circumcircle of any triangle in the Delaunay triangulation is empty (contains no nodes).
3. Empty circle property: Two points P_i and P_j are connected by an edge in the Delaunay triangulation, if and only if there is an empty circle passing through P_i and P_j.
4. Closest pair property: The closest pair of nodes in P are neighbors in the Delaunay triangulation. The circle having these two sites as its diameter cannot contain any other sites, and so is an empty circle.

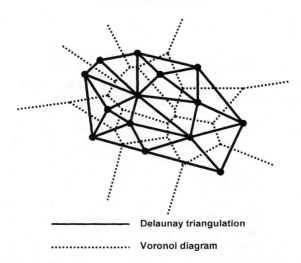

———————— Delaunay triangulation

···················· Voronoi diagram

Figure 3.5 The relationship between a Voronoi diagram and the Delaunay triangulation.

3.8 CONSTRUCTION OF TINS

The TIN model was developed in the 1970s as a simple way to build a surface from a set of irregularly spaced points (Peucker et al., 1978) as a method for avoiding redundancies in storing elevation data. Since then several prototype systems were developed in the 1970s based on their work. Commercial systems using TINs began to appear in the 1980s as contouring packages with some embedded in GIS. Several methods and techniques have emerged for TIN generation since then (McCullagh and Ross, 1980; Watson, 1981; and Mirante and Weingarten, 1982). Most of the developments were in the vector-based domain (de Berg et al., 2000). This section takes a closer look at the geometric construction of a TIN model based on Delaunay networks. There are several algorithms for the generation of Delaunay networks based on the source data (either grid or irregularly distributed points).

3.8.1 Creating TINs from Grid Data

Creating a TIN from grid data requires a number of decisions. The main three decisions are how to pick the significant points, how to connect these points into triangles, and how to model the surface of each individual triangle. A "significant point" refers to those points that are most useful in describing the surface and bringing out its salient topological features such as sharp variations. These points are then triangulated using triangulation algorithms to make the TIN model of the terrain.

Several algorithms have been proposed to select the "significant" points from the grid with minimal loss of information about the terrain. These methods differ in the criteria used to select the points in the DEM. Among these methods are the following:

Fowler and Little (1979) algorithm: The Fowler and Little algorithm defines significant points as the points that represent significant characteristics of the terrain such as peaks, pits, ridge lines, and channel lines. This algorithm works in an iterative manner; it requires a number of passes before the final selection of the "important" elevation points. The algorithm works on a 3×3 neighborhood and identifies peaks and pits (cells where all neighbors are either higher or lower in elevation). These points are then stored for a second processing cycle. The next processing cycle uses a 2×2 neighborhood to identify potential ridges and channels. A cell is a potential ridge point if it is higher than the three neighbors; it is a potential channel if it is lower than the three neighbors. A search is then conducted along these potential ridges toward the peaks and along the postenial channels toward pits. Peaks are connected to ridges while channels are connected to pits.

The Fowler and Little algorithm is suitable for only certain types of landscapes. It works well for landscapes with many sharp breaks of slopes, ridges, and sharp channels. The algorithm may not work well for an urban setting, where there are not many peaks and pits or ridges. The DEM can only be roughly approximated using Fowler and Little's algorithm. Moreover, the error in approximating the grid by a TIN would be different for different kinds of landscapes. In other words, the Fowler and Little algorithm cannot be used for TIN modeling where a maximum error has to be maintained. So this approach is not suitable for propagation prediction, especially for urban and semiurban areas since it can cause significant errors while doing propagation prediction for certain types of landscapes.

Very important points (VIP) algorithm (Chen and Guevara, 1987): In many cases, sample points for a DTM are redundant (i.e., many points only contribute in a very limited way to describe the surface). For example, in Figure 3.6 points A and B are located in the middle of the slope or in a flat area and therefore can be removed without losing any information about the surface. Some points, however, are more important since they indicate the "turning points" of the landform, such as ridges, basins, valleys, slope change points (e.g., points C and D in Figure 3.6). These "turning points" are called very important points (VIPs), which have the following characteristics:

1.   The slope at a VIP is discontinuous. In other words, a change of slope angle or aspect is expected at a VIP.

2.   If the sample point at a VIP is removed, then there will be significant loss of details about the surface.

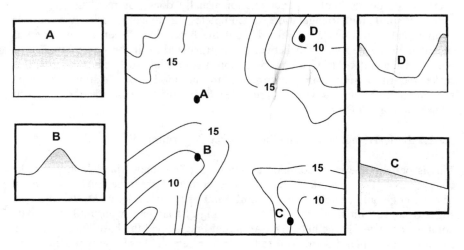

Figure 3.6 Very important points in a landscape.

The VIP algorithm assigns a certain "measure of significance" for all the points in the DTM based on the difference in elevation of the pixels with its neighbors. All the points with a measure of significance below a certain threshold are ignored while triangulating the DTM. The short description is that the VIP algorithm works to eliminate unimportant points. The method uses a 3×3 window and checks whether the central point offers any information that is different from its neighboring points. For more details, see Chen and Guevara (1987).

VIP gives a more accurate representation of the terrain when compared to the Fowler and Little algorithm. However, even this algorithm would not ensure that the maximum error of a DEM dataset, when represented using a TIN model, is below a certain value. Therefore, even though this method may give a more accurate representation of the terrain in most places, it may cause significant errors in some other regions.

3.8.2 Creating TINs from Irregularly Distributed Data

This section describes some algorithms for the generation of a TIN based on Delaunay triangulation. When the measured data points are well adapted to the terrain, the mathematical rules for Delaunay triangulation result in a triangular network well adapted for the terrain surface.

Many algorithms have been developed to achieve a Delaunay triangulation. These algorithms can, in general, be divided into two groups: static and dynamic triangulation. Static triangulation means that the triangulation is not valid before every point from the dataset is included in the network. There can be no qualified selection of points, and the triangulation usually does not meet the Delaunay criterion until the triangulation is complete. In dynamic triangulation the geometrical conditions are fulfilled from the beginning. When a new point is included in the network, the network is reorganized until the circle criterion is met. Consequently, during the triangulation it is possible to select the points that make the most significant contribution to the model. An example of both triangulation techniques will be presented here; for more details the reader is advised to consult Bjørke (1988).

3.8.2.1 Static Triangulation: The Step-by-Step Algorithm

This is perhaps the most common algorithm for Delaunay triangulation. For the construction of a network, it starts in one corner of the area, makes an initial triangle edge, and searches for the third point of the triangle. Next, the two new edges become "initial edges," and the building of triangles swell until every data point is connected in the network. An initial line, named the base line, is chosen. No other point should be interior to the circle where the base line is the diameter. This criterion is met by using two endpoints that are closest neighbors. If possible, the base line is situated on or close to the boundary of the area that is going to be triangulated.

The Delaunay neighbor of the line has to be found. It is clear that this neighbor, together with the two endpoints of the base line, defines a circle. None of the additional points, together with the endpoints, can define a circle with a lesser radius. The neighboring point can be found by calculating the angle α for the candidate points (Figure 3.7). The point that forms the largest α is the Delaunay neighbor of the line.

Lines between the new point and the two initial ones are stored as the second and third triangle edge. One triangle is now completed. The two new triangle edges from the last step become new base lines for further triangulation. The algorithm, then, successively builds a triangular network for the entire area. During the triangulation, only one side of the base line is searched for the neighboring point. The other side of the base line represents the triangulated area. For other static triangulation techniques, see McCullagh and Ross (1980) and Bjørke (1991).

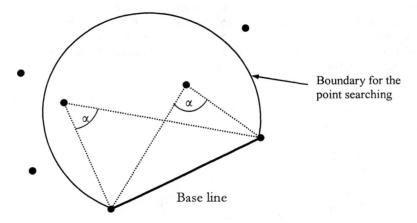

Boundary for the
point searching

Base line

Figure 3.7 Selection of the Delaunay neighbor of a line.

3.8.2.2 Dynamic Triangulation: The Incremental Algorithm

This algorithm was presented by Lee and Schachter (1980) and Guibas and Stolfi (1985). Further, de Floriani et al. (1985) used the algorithm for surface approximation. The most important aspect of the incremental algorithm is the possibility of keeping the triangular network as a Delaunay network during the triangulation process. The "initial value" for the triangulation is a valid Delaunay network (at least one triangle). For each point included in the network, the network will be rearranged until the circle criterion is met for all the triangles in the network. The general description of the incremental algorithm described below is given in more detail in Bjørke (1991).

As in the previous algorithm, an initial triangular network has to be created. The triangular network of the convex hull is used for this purpose [Figure 3.8(a)]. This network has to meet the maximum angle-sum criterion. The first point of the interior area is included in the network. The point is connected to its enclosing triangle by three new triangle edges between the point and the vertices of the triangle [Figure 3.8(b)].

The quadrilaterals, which have the "old" edges of the enclosing triangle as a diagonal, have to be tested by the maximum angle-sum rule. If they do not meet the criterion, their diagonals are swapped and the new, opposite edges to the inserted point will be examined as diagonals in their quadrilaterals. The results of the swapping operations are shown in Figure 3.8(c). The Delaunay network now contains one more point. All the remaining points will be included in the network in exactly the same way [Figure 3.8(d)].

The great advantage of this algorithm compared to the previous ones is the possibility of examination and calculation to determine whether any points make a significant contribution to the network. A plane is defined by three points only. An inclusion of a fourth point in the plane will not result in any improvements to the surface representation.

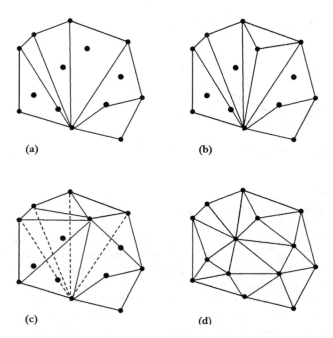

(a)   (b)

(c)   (d)

Figure 3.8 (a) An initial network. (b) Insertion of the first point. (c) The inconvenient edges are swapped. (d) Final lattice after the last point is included. (After Bjørke, 1988.)

3.9 CHOOSING THE APPROPRIATE MODEL

Source terrain data typically includes (1) irregularly spaced points (e.g., spot height points obtained from IfSAR or LIDAR systems), (2) contours, and (3) structure lines (or break lines) that capture the discontinuity of terrain and other important geographic features. Because a collection of individual points, contours, and break lines does not constitute a good (continuous) terrain representation in a digital environment (Peng et al., 1996), they are not usually used directly for surface visualization and analysis in geographic information systems (GIS). Instead, a typical GIS would build a DTM using this data, and carry out analysis based on the DTM. Because of this, terrain data is often stored and manipulated directly as a DTM, disregarding the source data. Furthermore, mapping projects often need to store and manage large amounts of digital terrain data. Even small projects may have to deal with a large amount of terrain data due to newly available data acquisition techniques such as LIDAR. Such data can be several terabytes in size, or may contain billions of measurement points. While most of today's GIS software are capable of storing and handling large volumes of spatial data, the question often arises as to which model is better, grid or TIN data models.

At this point it is not possible to answer this question in an unequivocal way. A short answer is neither and both. Neither of the two classes of data models is better in all conditions or for all data. Both have advantages and disadvantages relative to each other and to additional, more complex data models. In some instances it is preferable to maintain data in a grid model and in other cases in a TIN model. Most data may be represented in both, and may be converted among data models. The choice often depends on a number of factors, including the predominant type of data (discrete or continuous), the expected types of analyses, available storage, the main sources of input data, and the expertise of the human operator.

Many large data providers (USGS, for instance) choose the grid form for their terrain data due to its simplicity and relatively small storage size. TINs are typically used in places where engineering precision is required. Because of its sophisticated structure and heavy storage overhead (in order to keep topology), TINs are rarely used to provide and maintain a large amount of terrain data. A hybrid system that uses both grids and TINs may sound like a good solution, if it does not increase the complexity and difficulty in data management and updating, as well as in determining when to use the grid and when to use the TIN. Table 3.4 compares the characteristics of the two models. For further details, see Kumler (1994).

Table 3.4

A Comparison of Grid and TIN Data Models

Characteristics	Grid	TIN
Structure	Simple and easy to store and manipulate Connectivity is defined by neighboring cells	Complex and varies from one software to another Connectivity is based on the point-line-triangle topology definition
Georeferencing	Implicit in the grid structure	Implicit in the points coordinates
Storage requirements	Large if compression techniques are not implemented	More compact data storage, particularly for discrete objects
Data analysis	Analytical algorithms are easy to write Preferred for other raster layers combinations (e.g., satellite imagery)	Spatial operations are more complex Preferred for network analysis
Suitability for projects	Projects involving data with high spatial variability Projects in which original data is raster (e.g., satellites)	Projects requiring high precision of stored data Projects in which attributes are primarily character data
Advantages	Many spatial analysis functions often simpler and faster Efficient for data with high spatial variability Efficient for low spatial variability when compressed Easy to integrate with satellite and remotely sensed data	Linear geographic features such as streams and ridges are more accurately represented in a TIN Less points are needed to represent the topography Gives much greater precision and accuracy Greater flexibility in storing and manipulating attribute data
Disadvantages	Does not conform to variability of the terrain Linear features not well represented	Require visual inspection and manual control of the derived model
Data presentation	Full raster encoding (no compression) Run-length encoding Value-point encoding Quad trees Tiling (access strategy)	Point model Edge model Triangles model
Data availability	Grid data is available at both national and global levels (data is not expensive to obtain)	Point data is generally generated during data capturing

References

Bjørke, J.T., *Digital Kartografi.*, Department of Surveying and Mapping, Norwegian Institute of Technology, University of Trondheim, Part 3, 1991.

Bjørke, J.T., "Quadtrees and triangulation in digital elevation models," *International Archives of Photogrammetry and Remote Sensing, International Society for Photogrammetry and Remote Sensing, Committee of the 16th International Congress of ISPRS*, 27, part B4, Commission IV: 38-44, 1988.

Chen, Z., and Guevara, J.A., "Systematic selection of very important points (VIP) from digital terrain models for construction triangular irregular networks," *Proceedings, Auto-Carto 8*, ASPRS/ACSM, Falls Church, VA, 50-56, 1987.

Chen, Z.T., and Tobler, W., "Quadtree representations of digital terrain," *Proceedings, Auto-Carto* London, 1, 1986.

de Berg, M., van Kreveld, M., Overmars, M., and Schwarzkopf, O., *Computational Geometry, Algorithms and Applications*, 2nd ed., New York: Springer, 2000.

Defense Mapping Agency, *Digitizing the Future*, DMA Stock No. DDIPDIGITALPAC, Defense Mapping Agency, Fairfax, VA, 1990.

de Floriani, L., Falcidieno, B., and Pienovi, C., "Delaunay-based representation of surfaces defined over arbitrarily shaped domains," *Computer Vision, Graphics, and Image Processing*, 32, 1985.

Delaunay, B., "Sur la sphere vide," *Bulletin of Academy of Sciences of the USSR:* 793-800, 1934.

Dirichlet, G.L., "Über die reduction der positiven quadratischen formen mit drei unbestimmten ganzen zalen," *J. Reine u. Angew. Math.*, 40:209-227, 1850.

Fowler, R.J., and Little, J.J., "Automatic extraction of irregular network digital terrain models," *ACM*, 4:199-200, 1979.

Gaustad, A., "Generalisering av 2.5d-geografiske flater," Master's thesis, Department of Surveying and Mapping, Norwegian Institute of Technology, University of Trondheim, 7034, Trondheim, Dec. 1990.

Guibas, L., and Stolfi, J., "Primitives for the manipulation of general subdivisions and the computation of Voronoi diagrams," *ACM Transactions on Graphics*, 4(2): 74-123, 1985.

Heller, M., "Triangulation algorithms for adaptive terrain modeling," *Proceedings of the 4th International Symposium on Spatial Data Handling,* 163-174, July 1990.

Kumler, M.P., "An intensive comparison of triangulated irregular networks (TINs) and digital elevation models," *Cartographica*, 31:1-99, 1994.

Lee, D.T., and Schachter, B.J., "Two algorithms for constructing a Delaunay triangulation," *International Journal of Computer and Information Sciences*, 9(3): 219-242, 1980.

McCullagh, M.J., and Ross, C.G., "Delaunay triangulation of a random data set for isarithmic mapping," *The Cartographic Journal*, 17(2): 93-99, 1980.

Mirante, A., and Weingarten, N., "The radial sweep algorithm for constructing triangulated irregular networks," *IEEE Computers Graphics & Applications*, 2: 11-21, 1982.

Palacios-Velez, O.L., and Cuevas-Renaud, B., "Automated river-course, ridge and basin delineation from digital elevation data," *Journal of Hydrology*, 86: 299-314, 1986.

Peng, W., Petrovic, D., and Crawford, C., "Handling large terrain data in GIS," *Com 4, ISPRS Congress*, Istanbul, 2004.

Peng, W., Pilouk, M., and Tempfli, K., "Generalizing relief representation using digital contours," *Proceedings of the XVIII ISPRS Congress*, Vienna, Austria: 8, July 9-19, 1996.

Peucker, T.K., Fowler, R.J., Little, J.J., and Mark, D.M., "The triangulated irregular network," *Proceedings, American Society of Photogrammetry: Digital Terrain Models (DTM) Symposium*, St. Louis, MO, 516-540, May 9-11, 1978.

Samet, H., "The quadtree and related data structures," *ACM Computing Surveys*, 16(2):187–260, 1984.

Tucker, G. E., and Slinglerland, R.L., "Drainage basin response to climate change," *Water Resources Research*, 33(8): 2031-2047, 1997.

U.S. Geological Survey, *Catalog of U.S. GeoData*, 1990.

Voronoi, M.G., "Nouvelles applications des parametres continus a la theorie des formes quadratiques," *J. Reine u. Angew. Math.*, 134:198-287, 1908.

Watson, D.F., "Computing the n-dimensional Delaunay tessellation with applications to Voronoi polytypes," *The Computer Journal*, 24:167-172, 1981.

Chapter 4

DTM Manipulation

4.1 INTRODUCTION

The acquisition of DTM data aims at representing the continuous surface of the Earth using a discrete set of points, which could be handled by computers. In other words, DTM acquisition is a sampling process that represents a continuous surface through a finite set of points. It should be noted that alternative sampling schemes may lead to different results both in the DTM itself and its derivatives (e.g., slope maps, aspect maps, and contour lines). Therefore, the accuracy of the individual elevations/samples is not the only criterion that controls the accuracy of derived information from a DTM. Other contributing factors include the density and the locations of the sample points relative to key surface characteristics, such as break lines, mountain peaks, and valley basins. DTM acquisition techniques have been discussed in Chapter 2.

DTM manipulation includes editing, filtering, enhancing, merging, joining, and elevation resampling. Editing the DTM data might be required to correct errors in the elevation data resulting from erroneous matches in overlapping images and/or to update portions of the DTM due to object space changes (e.g., construction of new roads, cut and fill applications). DTM filtering and/or enhancement might be used to highlight the trend or the local detail component of the surface in question. In addition, a filtering procedure might be carried out to reduce the data volume in a DTM, which in turn leads to saving disk storage space and/or efficient data processing. Merging and joining DTM data is necessary when dealing with multiple datasets, which have been captured by the same or different sensors at different times. Finally, DTM resampling aims at reconstructing the continuous surface of the Earth from sampled elevations. In other words, the resampling procedure tries to invert the sampling process. Therefore, DTM resampling is usually known as surface representation from point data or simply interpolation. Among DTM manipulation activities, resampling is the most important and frequently used procedure, and is the focus of this chapter.

4.2 SURFACE CLASSIFICATION

Before getting into the details of DTM interpolation techniques, let us start by discussing various surface classifications in terms of representation, continuity, and smoothness. These classification alternatives will be called upon when discussing various interpolation techniques.

4.2.1 Functional Versus Solid Surfaces

Functional surfaces are only capable of storing a single Z value for a given (X, Y) location. The most common example of a functional surface is the terrain along the Earth's surface. Other examples of terrestrial functional surfaces include bathymetric data and an underground water table. Functional surfaces can also be used to represent statistical surfaces describing climatic and demographic data, concentration of resources, and other biological data. In addition, functional surfaces can be used to represent mathematical surfaces as defined by arithmetic expressions such as $Z = a + bX + cY$, which denotes a planar surface. The term *2.5-dimensional surface* has been frequently used to denote functional surfaces.

Functional surface models can be contrasted to *solid surfaces*, which are true three-dimensional models capable of storing multiple Z values for a given (X, Y) location. Solid models are common in computer assisted design (CAD) and engineering applications. Examples of objects suited to solid modeling include machine parts, buildings, and other objects placed on or below the Earth's surface (e.g., buildings, bridges, and tunnels). In some cases, it is possible to represent some three-dimensional objects such as faults and buildings using functional surfaces by slightly offsetting the duplicate (X, Y) coordinates.

4.2.2 Continuous Versus Noncontinuous Surfaces

Functional surfaces are known to be continuous. That is, when approaching a given (X, Y) location on a functional surface from any direction, one will get the same Z value. This can be contrasted to *discontinuous surfaces*, where different Z values can be obtained depending on the direction of approach; see Figure 4.1. An example of a discontinuous surface is a vertical fault across the surface of the Earth. A location at the top of a fault has one elevation, while immediately below this point at the bottom of the fault, one observes another elevation. A model capable of handling a discontinuous surface should allow for storing more than one Z value for a given (X, Y) location.

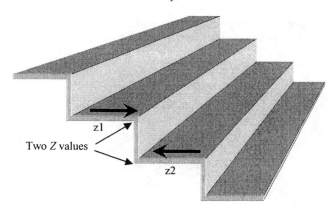

Two Z values

Figure 4.1 Discontinuous surface.

4.2.3 Smooth Versus Nonsmooth Surfaces

The smoothness property deals with the change of the surface normal from one location to the next. The surface normal is defined as the normal vector to the surface's first derivative (slope). In addition to being continuous, a *smooth surface* has the property that regardless of the direction from which one approaches a given point on the surface, the direction of the normal vector will be the same.

Illustrations of two surfaces with different levels of smoothness are shown in Figures 4.2 and 4.3. The first surface, represented by two planar facets, is not smooth. Within each planar facet, the surface normal is constant. However, as one moves across the intersection of the two facets, the surface normal abruptly changes. In contrast to Figure 4.2, a smooth surface is shown in Figure 4.3, where the surface normal varies continuously across the surface. Terrain surfaces vary in smoothness. For example, geologically young terrain typically has sharp ridges and valleys. On the other hand, older terrain is usually smoothed by prolonged exposure to erosion forces and weathering conditions. Statistical surfaces such as those representing rainfall or air temperature are generally smooth.

4.3 SURFACE INTERPOLATION: INTRODUCTION

Having discussed the various alternatives for describing surface characteristics, the rest of this chapter will focus on the discussion of surface resampling/interpolation techniques. As mentioned earlier, those techniques aim at reconstructing the surface of the Earth using a discrete set of elevations. In general, *interpolation* can be defined as the process of estimating the value of attributes at some sites from measurements made at surrounding point locations, which are denoted as *sample/reference points*. This is in contrast to *extrapolation*,

which is the process of predicting the value of an attribute at sites outside the area
covered by existing reference points; see Figure 4.4.

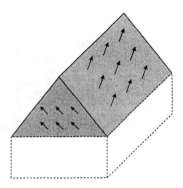

Figure 4.2 The direction of the surface normal abruptly changes along a nonsmooth surface.

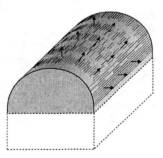

Figure 4.3 The surface normal along a smooth surface does not abruptly change.

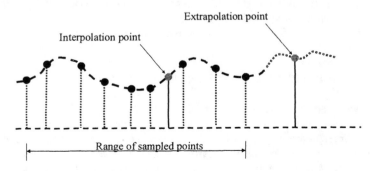

Figure 4.4 Interpolation versus extrapolation.

4.3.1 Motivation

Interpolation is usually used for surface representation from point data. For DTM applications, the interpolation process might be needed for one or more of the following reasons:

1. The sampled surface has a different level of resolution (e.g., cell size) and/or orientation from that required. For example, interpolation can be used to convert a 1-m grid aligned along the east and north directions into a 2-m grid aligned along the northeast and northwest directions.
2. The sampled elevations need to be transformed from one format to another. For example, the interpolation can be used to convert an irregular set of samples into a regular one along a grid.
3. The need to convert point data to surface representation, which can be used with other surfaces and/or data sources for analysis and modeling (e.g., change detection applications, orthophoto generation, and assessment of fire risk).

The above needs might be required for DTM applications related to civil engineering, earth science, planning and resource management, surveying, photogrammetry, environment, and defense. In general, the interpolation process might involve one or more of the following operations:

1. Compute the elevation (Z) at distinct point locations.
2. Compute the elevation (Z) at a rectangular grid from irregularly sampled points (so-called *gridding*).
3. Compute the (XY) locations of points along contours (locus of points with the same elevation).
4. Densification or coarsening of grids.

There is abundant literature related to DTM interpolation methods, some of which will be discussed in the remainder of this chapter. However, before getting into the mathematical details of these interpolation methods, one should note the following characteristics and peculiarities of DTM interpolation:

1. There is no "best" interpolation algorithm that is clearly superior to all others and appropriate for all applications.
2. The quality of the resulting DTM is controlled by the distribution and accuracy of the original sample/reference points as well as the adequacy of the underlying assumptions of the interpolation model (i.e., a hypothesis about the behavior of the terrain surface).
3. The most important criterion for selecting a DTM interpolation method is the degree to which the key surface features (e.g., break lines) can be taken into account and its capability in adapting to varying terrain characteristics.

4.3.2 Classification of Interpolation Methods

Interpolation methods can be classified according to the following criteria:

1. The compatibility between the interpolated elevations at the sampled points and the original elevations (*exact and inexact interpolation methods*).
2. The spatial extent of the utilized samples for the estimation of the elevation at a given interpolation point (*global and local interpolation methods*).
3. The utilized terrain and data characteristics within the interpolation mechanism (*stochastic and deterministic interpolation methods*).

The following paragraphs briefly outline the classification of interpolation methods according to the above criteria:

Exact Versus Inexact Interpolation: An exact interpolation estimates an elevation at a reference point location that is the same as the given elevation at that point. In other words, exact interpolation generates a surface that passes through the reference points. An interpolation method, which yields a different elevation value at the reference points when compared to the given elevations at these points, is known as an inexact interpolator. In this case, the statistical differences (absolute and/or squared) between the given and interpolated elevations at the sample points are often used as indicators of the quality of an inexact interpolator. Figures 4.5 (a,b) illustrate the outcome from exact and inexact interpolators, respectively.

Global Versus Local Interpolation: The interpolation methods can be classified into "global" and "local" techniques depending on the portion of the utilized samples to estimate the elevation at a given point. As the name suggests, global interpolation uses all the available samples to estimate the elevation at the interpolation point. In contrast, local interpolation estimates the unknown elevation using these associated with the nearest sample points.

Stochastic Versus Deterministic Interpolation: Deterministic methods do not consider the statistical properties of the surface and the sampled elevations within the interpolation mechanism. On the other hand, stochastic approaches consider the statistical properties of the surface and the sampled elevations at the reference points throughout the interpolation procedure.

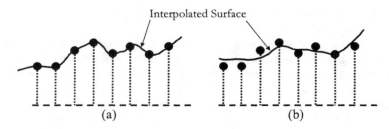

Figure 4.5 (a) Exact and (b) inexact interpolation methods.

4.4 INTERPOLATION METHODS

The following sections present the mathematical and conceptual details of global/local, exact/inexact, and deterministic/stochastic interpolation methods. Throughout these discussions, we will be using $Z_i(X_i, Y_i)$ to denote the measured elevation at the ith reference point whose horizontal coordinates are (X_i, Y_i). On the other hand, $\hat{Z}_i(X_i, Y_i)$ denotes the estimated elevation at the interpolation point whose coordinates are (X_i, Y_i).

4.4.1 Global Interpolation Methods

As mentioned earlier, a global interpolation method uses all the known elevations at the reference points to estimate the unknown elevation at the interpolation point. The following sections provide the technical details of some of the global interpolation methods such as trend surface analysis (TSA), Fourier analysis, and KRIGING.

4.4.1.1 Trend Surface Analysis

Trend surface analysis (TSA) is the most widely used global interpolation procedure. TSA is a deterministic interpolation, which approximates the surface by a polynomial expansion of the geographic coordinates (i.e., XY-coordinates). In other words, an observed elevation at a reference point is considered to be the sum of a polynomial function of its XY-coordinates (trend) and a random error (residual); see Figure 4.6. The involved coefficients in the polynomial function are estimated through a least squares adjustment procedure using the measured elevations at the reference points. The least squares adjustment will minimize the squared sum of deviations/residuals between the original surface and derived trend from the polynomial expansion. Having estimated the polynomial coefficients, it is a simple matter to estimate the elevation at any location (X, Y). For example, one can create a grid of elevations by substituting the coordinates of the grid nodes into the polynomial and calculating the elevation at each node. Theoretically, one can choose a polynomial of any order to represent the trend as long as there are a sufficient number of reference points. However, there will be some limitations resulting from the exponential growth of the numerical values. Also, the estimated coefficients might not be representative of the surface in question. The TSA mathematical details and the relevance of the chosen polynomial will be discussed in the following sections.

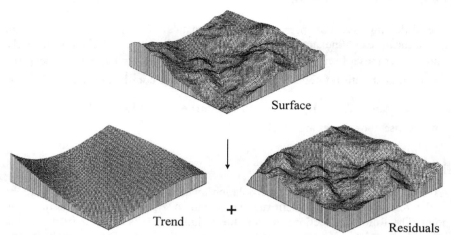

Surface

Trend

+

Residuals

Figure 4.6 Principle of trend surface analysis: the original surface is considered to be the summation of a trend and residuals.

TSA Mathematics

As mentioned earlier, TSA defines a surface as the summation of a trend, which is described by a deterministic polynomial expansion and residuals. The trend describes the global characteristics of the surface while the residuals can be viewed as the local surface details. For example, (4.1) expresses the measured elevation at a given reference point as the summation of polynomial expansion and a residual e:

$$Z(X, Y) = a_0 + a_1 X + a_2 Y + \ldots\ldots + e \tag{4.1}$$

Where $Z(X, Y)$ is the measured elevation at the reference point, $(a_0, a_1, a_2, \ldots..)$ are the polynomial coefficients, and e is the corresponding residual. The optimum trend is the one that minimizes the squared sum of residuals (i.e., $\sum e^2 =$ minimum). This trend is achieved through the use of a least squares adjustment (Koch, 1987). The order of the polynomial that represents the trend is pre-specified by the user. The general form of a trend, which is described by an nth order polynomial expansion, is given in (4.2).

$$Z(X, Y) = \sum_{i=0}^{n} \sum_{j=0}^{n-i} a_{ij} X^i Y^j \tag{4.2}$$

For the purpose of illustration, let us consider the one-dimensional profile in Figure 4.7(a). This profile can be separated into a trend and local fluctuation components in a variety of ways [e.g., first order, Figure 4.7(b); second order, Figure 4.7(c); or third order, Figure 4.7(d)].

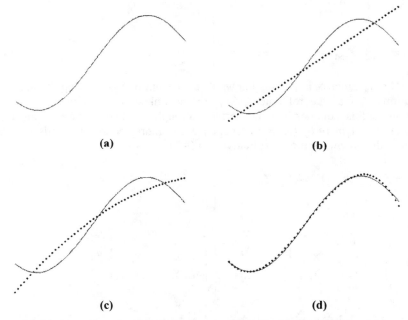

(a)　　　　　　　　　　　　　　(b)

(c)　　　　　　　　　　　　　　(d)

Figure 4.7 (a) A one-dimensional profile can be approximated by (b) a first-order trend, (c) second-order trend or (d) third-order trend.

To illustrate the least squares procedure for estimating the polynomial coefficients, let us consider a first-order trend, which will be fitted to *n* reference points. In this case, the measured elevations at the reference points will be mathematically related to the unknown coefficients as in (4.3):

$$Z_1(X_1, Y_1) = a_0 + a_1X_1 + a_2Y_1 + e_1$$
$$Z_2(X_2, Y_2) = a_0 + a_1X_2 + a_2Y_2 + e_2$$
$$\vdots \hspace{9cm} (4.3)$$
$$Z_n(X_n, Y_n) = a_0 + a_1X_n + a_2Y_n + e_n$$

Equation (4.3) can be expressed in a matrix form as in (4.4), or as the symbolic form in (4.5).

$$
\begin{bmatrix} Z_1 \\ Z_2 \\ \vdots \\ Z_n \end{bmatrix}_{n \times 1} = \begin{bmatrix} 1 & X_1 & Y_1 \\ 1 & X_2 & Y_2 \\ \vdots & \vdots & \vdots \\ 1 & X_n & Y_n \end{bmatrix}_{n \times 3} \begin{bmatrix} a_0 \\ a_1 \\ a_2 \end{bmatrix}_{3 \times 1} + \begin{bmatrix} e_1 \\ e_2 \\ \vdots \\ e_n \end{bmatrix}_{n \times 1}
\tag{4.4}
$$

$$
Z_{n \times 1} = A_{n \times 3} \, X_{3 \times 1} + e_{n \times 1}
\tag{4.5}
$$

The $A_{n \times 3}$ matrix in (4.5) is known as the design matrix relating the observed elevations $Z_{n \times 1}$ at the reference points to the unknown coefficients $X_{3 \times 1}$. The solution to this equation can be derived through the least squares adjustment formulation as in (4.6) (refer to Chapter 5 for more technical details on least squares adjustment) or more explicitly as in (4.7).

$$
\hat{X}_{3 \times 1} = (A_{3 \times n}^T \, A_{n \times 3})^{-1} \, A_{3 \times n}^T \, Z_{n \times 1}
\tag{4.6}
$$

$$
\begin{bmatrix} \hat{a}_0 \\ \hat{a}_1 \\ \hat{a}_2 \end{bmatrix}_{3 \times 1} = \left(\begin{bmatrix} 1 & 1 & \cdots & 1 \\ X_1 & X_2 & \cdots & X_n \\ Y_1 & Y_2 & \cdots & Y_n \end{bmatrix}_{3 \times n} \begin{bmatrix} 1 & X_1 & Y_1 \\ 1 & X_2 & Y_2 \\ \vdots & \vdots & \vdots \\ 1 & X_n & Y_n \end{bmatrix}_{n \times 3} \right)^{-1}_{3 \times 3} \begin{bmatrix} 1 & 1 & \cdots & 1 \\ X_1 & X_2 & \cdots & X_n \\ Y_1 & Y_2 & \cdots & Y_n \end{bmatrix}_{3 \times n} \begin{bmatrix} Z_1 \\ Z_2 \\ \vdots \\ Z_n \end{bmatrix}_{n \times 1}
$$

$$
\begin{bmatrix} \hat{a}_0 \\ \hat{a}_1 \\ \hat{a}_2 \end{bmatrix}_{3 \times 1} = \begin{bmatrix} n & \sum_{i=1}^{n} X_i & \sum_{i=1}^{n} Y_i \\ \sum_{i=1}^{n} X_i & \sum_{i=1}^{n} X_i^2 & \sum_{i=1}^{n} X_i Y_i \\ \sum_{i=1}^{n} Y_i & \sum_{i=1}^{n} Y_i X_i & \sum_{i=1}^{n} Y_i^2 \end{bmatrix}_{3 \times 3}^{-1} \begin{bmatrix} \sum_{i=1}^{n} Z_i \\ \sum_{i=1}^{n} X_i Z_i \\ \sum_{i=1}^{n} Y_i Z_i \end{bmatrix}_{3 \times 1}
\tag{4.7}
$$

The solution to the above equations will yield the best-fit as defined by a least squares adjustment between the measured elevations at the reference points and a first-order (plane) trend surface. The inverse of the *normal matrix* $(A^T A)$ will not exist if we have less than three reference points. Therefore, the number of the reference points (n) should be larger than the number of coefficients in the selected trend polynomial. Examples of second-, third-, and fourth-order trend surfaces are shown in (4.8), (4.9), and (4.10), respectively.

$$
Z(X, Y) = a_0 + a_1 X + a_2 Y + a_3 X^2 + a_4 XY + a_5 Y^2
\tag{4.8}
$$

$$Z(X, Y) = a_0 + a_1 X + a_2 Y$$
$$+ a_3 X^2 + a_4 XY + a_5 Y^2 \tag{4.9}$$
$$+ a_6 X^3 + a_7 X^2 Y + a_8 XY^2 + a_9 Y^3$$

$$Z(X, Y) = a_0 + a_1 X + a_2 Y$$
$$+ a_3 X^2 + a_4 XY + a_5 Y^2$$
$$+ a_6 X^3 + a_7 X^2 Y + a_8 XY^2 + a_9 Y^3 \tag{4.10}$$
$$a_{10} X^4 + a_{11} X^3 Y + a_{12} X^2 Y^2 + a_{13} XY^3 + a_{14} Y^4$$

Having obtained the coefficients of the trend surface, we can estimate the elevation or trend value $\hat{Z}(X, Y)$ at any location (X, Y); refer to (4.11) for the case of a first-order trend.

$$\hat{Z}(X, Y) = \begin{bmatrix} 1 & X & Y \end{bmatrix} \begin{bmatrix} \hat{a}_0 \\ \hat{a}_1 \\ \hat{a}_2 \end{bmatrix} \tag{4.11}$$

Assuming that n reference points have been used to estimate the polynomial coefficients, one can rewrite (4.11) as follows:

$$\hat{Z}(X, Y) = \begin{bmatrix} 1 & X & Y \end{bmatrix} \left(\begin{bmatrix} 1 & 1 & \cdots & 1 \\ X_1 & X_2 & \cdots & X_n \\ Y_1 & Y_2 & \cdots & Y_n \end{bmatrix}_{3 \times n} \begin{bmatrix} 1 & X_1 & Y_1 \\ 1 & X_2 & Y_2 \\ \vdots & \vdots & \vdots \\ 1 & X_n & Y_n \end{bmatrix}_{n \times 3} \right)^{-1}_{3 \times 3} \begin{bmatrix} 1 & 1 & \cdots & 1 \\ X_1 & X_2 & \cdots & X_n \\ Y_1 & Y_2 & \cdots & Y_n \end{bmatrix}_{3 \times n} \begin{bmatrix} Z_1 \\ Z_2 \\ \vdots \\ Z_n \end{bmatrix}_{n \times 1}$$

$$\hat{Z}(X, Y) = \sum_{i=1}^{n} w_i Z_i \tag{4.12}$$

Equation (4.12) indicates that the estimated elevation at the interpolation point is a weighted average of the elevations at the reference points. The weighting function depends on the XY-coordinates of the interpolation and reference points. Now, the question to be asked is how can we judge whether the appropriate trend, polynomial order has been selected or not? The answer to this question will be provided through the TSA analysis of variance.

TSA Analysis of Variance

A given order trend surface can be fitted to any set of reference points, but that does not mean that it is a meaningful or worthwhile trend. One might think that it is always better to use a higher-order trend function rather than a lower-order one. In other words, it might be expected that as the polynomial order increases, the trend gets closer to the original surface. However, this is not always true. To evaluate the "worthiness" of the trend surface, we need to derive the total sum of squares (SS_T), the regression/trend sum of squares (SS_R), and the residual sum of squares (SS_D) as described in (4.13), (4.14), and (4.15), respectively.

$$SS_T = \sum_{i=1}^{n}(Z_i - \overline{Z})^2 = \sum_{i=1}^{n}Z_i^2 - (\sum_{i=1}^{n}Z_i)^2 \Big/ n \qquad (4.13)$$

$$SS_R = \sum_{i=1}^{n}(\hat{Z}_i - \overline{Z})^2 = \sum_{i=1}^{n}\hat{Z}_i^2 - (\sum_{i=1}^{n}\hat{Z}_i)^2 \Big/ n = \sum_{i=1}^{n}\hat{Z}_i^2 - (\sum_{i=1}^{n}Z_i)^2 \Big/ n \qquad (4.14)$$

$$SS_D = \sum_{i=1}^{n}(Z_i - \hat{Z}_i)^2 = \sum_{i=1}^{n}Z_i^2 - \sum_{i=1}^{n}\hat{Z}_i^2 = SS_T - SS_R \qquad (4.15)$$

The $(\overline{Z})$ in (4.13) and (4.14) represents the average elevation as defined by the reference points. Equations (4.14) and (4.15) assume that the trend and the original surface have the same average. Moreover, it is also assumed that there is no correlation between the estimated trend and the remaining residuals. These assumptions are valid since the residuals are assumed to represent local/random fluctuations, which should average to zero. Therefore, the trend and the original surface should have the same average. Also, the trend, which is the systematic component of the surface as defined by the polynomial expansion, should have no correlation with the residual/random component of the surface. Using the above sums of squares, one can define the percentage of goodness-of-fit of the trend (R^2) as in (4.16).

$$R^2 = \frac{SS_R}{SS_T} \qquad (4.16)$$

The equality in (4.15) $\left(SS_D = SS_T - SS_R\right)$ will guarantee that the percentage of goodness-of-fit (R^2) will be less than 1.0 (note that SS_D is always positive since it represents a sum of squares). The optimum value for (R^2) is 1.0, which indicates that ($SS_T = SS_R$). This will only happen if the estimated trend coincides with the original surface as defined by the reference points (i.e., we have an exact interpolator). Since it is almost impossible to have a terrain that can be exactly described by a deterministic polynomial, the value of (R^2) will never reach 1.0. Therefore, a good trend surface will yield a high percentage of goodness of fit. Thus, a strategy for determining the optimum trend can start with a low-order polynomial (e.g., first-order). Then, we increase the order of the polynomial while observing (R^2). If the value of R^2 is increasing as we increase the polynomial order, then the higher the polynomial order, the better is the fit. However, if one reaches a stage where increasing the order has a minor effect on the value of R^2, then it is better to use the lowest polynomial order while maintaining the value of R^2.

An alternative methodology for testing the significance of a trend surface is by performing an analysis of variance (ANOVA). The ANOVA starts by computing the variances MS_T, MS_R, and MS_D, which correspond to the total sum of squares (SS_T), the trend sum of squares (SS_R), and the residual sum of squares (SS_D), respectively. These variances are derived by dividing the sum of squares by the corresponding degrees of freedom; refer to (4.17), (4.18), and (4.19). The degrees of freedom associated with the total sum of squares SS_T is ($n - 1$), where n is the number of reference points. The degrees of freedom associated with the trend sum of squares SS_R is ($m - 1$), where m is the number of coefficients in the polynomial describing the trend. Finally, the degrees of freedom for the residual sum of squares SS_D is the difference between the previous two; that is, ($n - m$).

$$\text{Total variance} = MS_T = \frac{SS_T}{n-1} \qquad (4.17)$$

$$\text{Trend variance} = MS_R = \frac{SS_R}{m-1} \qquad (4.18)$$

$$\text{Residual variance} = MS_D = \frac{SS_D}{n-m} \qquad (4.19)$$

To test the significance of the estimated trend, we need to determine the ratio between the trend variance and the residual variance, (4.20). This ratio follows an F distribution with ($m - 1$, $n - m$) degrees of freedom (Rice, 1987).

Learning Resources
Centre

$$F_{statistic} = \frac{MS_R}{MS_D} \sim F_{(m-1,\, n-m)} \tag{4.20}$$

If the regression is significant (i.e., the trend is very close to the original surface), the residual variance MS_D will be very small when compared to the trend variance MS_R leading to a large $F_{statistic}$ value. Therefore, the $F_{statistic}$ can be used in an F-test to determine if the trend-surface coefficients are significantly different from zero (i.e., the regression effect is significantly different from the random effect of the data). In formal statistical terms, the null (H_O) and alternative (H_A) hypotheses within the F-test for the significance of the trend can be formulated as in (4.21):

$$H_O : a_0 = a_1 = a_2 = \ldots\ldots\ldots = a_{m-1} = 0$$
$$H_A : a_0, a_1, a_2, \ldots\ldots\ldots\ldots, a_{m-1} \neq 0 \tag{4.21}$$

The null hypothesis (H_O) suggests that the regression effect is not significantly different from the random effect of the data. Accepting the null hypothesis leads to the conclusion that some or all of the polynomial coefficients are not significant and the order of the polynomial can be reduced. On the other hand, if the null hypothesis (H_O) is rejected (i.e., the alternative hypothesis, H_A, is accepted), all the coefficients in the regression are considered to be significant and the regression is worthwhile. The value of the $F_{statistic}$, (4.20), can be checked against the critical values in the appropriate F-distribution tables while assuming a level of significance (α), which is the probability of rejecting a true null hypothesis – probability of type I error.

Numerical Example

Figure 4.8 shows a 3-D plot of 161 irregularly distributed reference points. These reference points are used in TSA analysis to derive trend surfaces that are defined by first-order, second-order, and third-order polynomial functions. The estimated coefficients of these polynomials are shown in Table 4.1. As an example, Figure 4.9 illustrates the second-order trend surface, as defined by the coefficients in the third column of Table 4.1. A closer look at Figures 4.8 and 4.9 indicates that the derived trend is smoother than the original data. The total variance (MS_T), trend variance (MS_R), residual variance (MS_D), goodness-of-fit (R^2), and the $F_{statistic}$ [together with the corresponding degrees of freedom (DOF)] for the different trend surfaces are listed in Table 4.2.

A closer look at the numerical values in Table 4.2 reveals the following facts:

1. The total sum of squares (SST) does not change as we increase the order of the trend surface polynomial. This should be expected since SST is a measure of the deviation between the original surface and its mean.
2. The trend sum of squares (SSR) gets larger as we increase the order of the trend surface polynomial. Such a finding should come as no surprise since the departure of the trend from its mean should increase with increasing the order of the trend surface polynomial.
3. The residual sum of squares (SSD) gets smaller as we increase the order of the trend surface polynomial. Once again, this should be expected since the trend surface should get closer to the original surface as we increase the order of the trend surface polynomial.
4. For the different trend surfaces, the total sum of squares (SST) is equivalent to the summation of the trend sum of squares (SSR) and the residual sum of squares (SSD); refer to (4.15). This is why an increase in the trend sum of squares (SSR) should be coupled with a reduction in the residual sum of squares (SSD).
5. The goodness-of-fit (R2) increases as we increase the order of the trend surface polynomial. Such an increase indicates a closer similarity between the trend and the original surface. However, one can see that the rate of increase between the second- and third-order trends is much smaller than that between the first- and second-order trends. Therefore, one should expect that further increase in the order of the trend polynomial will not necessarily lead to an improvement in the goodness of fit (i.e., at some stage, increasing the order of the trend polynomial function might be accompanied with insignificant improvement in the goodness of fit).
6. All the estimated trends can be considered to be worthwhile according to the derived $F_{statistic}$ in Table 4.2. More specifically, the derived $F_{statistic}$ is larger than the critical value, which is shown in the last row in Table 4.2. These critical values correspond to a level of significance (α) of 0.05.

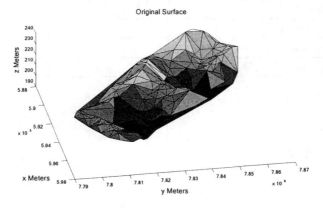

Figure 4.8 A 3-D plot of 161 irregularly distributed reference points.

Digital Terrain Modeling

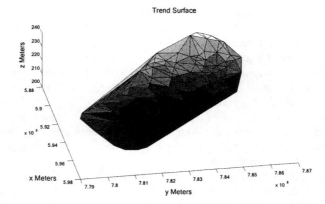

Figure 4.9 Second-order trend surface derived from the reference points in Figure 4.8.

Table 4.1

Coefficients of First-, Second-, and Third-Polynomial Functions Representing the Trend of the Reference Points in Figure 4.8

	First Order	*Second Order*	*Third Order*
Const.	1.09576118	1.34899455	0.20045373
X Coefficient	-0.81875923	-1.90417990	1.29890560
Y Coefficient	-0.43639678	-0.62377634	4.11695850
X^2 Coefficient	0.0	1.06733263	-2.28741957
XY Coefficient	0.0	-0.00927066	-6.67368031
Y^2 Coefficient	0.0	0.14646641	-6.65967351
X^3 Coefficient	0.0	0.0	1.32759548
X^2Y Coefficient	0.0	0.0	2.38235048
XY^2 Coefficient	0.0	0.0	4.11002662
Y^3 Coefficient	0.0	0.0	3.22251964

Table 4.2

TSA Analysis of Variance for the Derived First-, Second-, and Third-Trend Surfaces in Table 4.1

	First Order	*Second Order*	*Third Order*
SST/DOF (MST)	10947/160 (68)	10947/160 (68)	10947/160 (68)
SSR/DOF (MSR)	4764/2 (2382)	6193/5 (1239)	6425/9 (714)
SSD/DOF (MSD)	6183/158 (39)	4754/155 (31)	4522/151 (30)
R2	0.43520567	0.56574501	0.58693519
F-Statistic (DOF) (Pass/Fail)	60.8739259 (2,158) Pass	40.3866286 (5,155) Pass	23.8400072 (9,151) Pass
F (critical value) $\alpha = 0.05$	3.0	2.21	1.88

TSA: Concluding Remarks

The TSA interpolation procedure has the advantages of deriving a unique surface (e.g., the same surface is generated regardless of the chosen reference frame for the sample/reference points), being easy to program, and requiring relatively low computational power, especially when using a low-order trend. On the negative side, the assumption of having a surface that is modeled by a deterministic polynomial is rarely met in practice. In some cases, the surface might be too complicated to be represented by a low-order polynomial. Also, it is difficult to justify the use of a high-order polynomial to describe the trend. The derivation of a reliable trend requires having adequate reference points. The number of reference points should be much greater than the number of unknown polynomial coefficients. The spacing and the distribution of the reference points are crucial for ensuring high quality of the derived trend. Clusters and/or uneven distribution of the reference points will affect the shape of the estimated surface, leading to problems or biases. In general, one might encounter unreasonable trend behavior near the surface boundaries. The trend can overshoot and/or undershoot at the boundaries. Therefore, it is important to have a buffer around the area of concern to avoid these problems. Finally, to avoid numerical instabilities when inverting the normal equation matrix, it is recommended to normalize the coordinates prior to the interpolation process.

4.4.1.2 Fourier Analysis

Fourier analysis is generally used to decompose spatial or time-domain signals as the summation of scaled basis functions in the form of complex exponentials/sinusoidal waves with different frequencies (*f*). Signal decomposition as the summation of sinusoidal functions is usually known as transforming the signal into the frequency domain (*forward Fourier transform*). An example of forward and backward Fourier transforms of continuous one-dimensional signals between the spatial and frequency domain is illustrated in (4.22) and (4.23), respectively (Oppenheim and Schafer, 1989). Equation (4.22) is used to derive the frequency-domain representation of the signal $G(f)$, which is known as the *spectrum*, from the corresponding spatial-domain signal $g(x)$. On the other hand, (4.23) derives the spatial function $g(x)$ from its spectrum $G(f)$.

$$G(f) = \int_{-\infty}^{\infty} g(x)\, e^{-i2\pi fx}\, dx \qquad\qquad (4.22)$$

$$g(x) = \int_{-\infty}^{\infty} G(f)\, e^{i2\pi fx}\, df \qquad\qquad (4.23)$$

In DTM applications, one can consider the terrain as a continuous two-dimensional signal, which can be transformed into the frequency domain using the Fourier transform. Due to the relatively smooth nature of the terrain, one expects that the amplitudes $G(f)$ of high-frequency components of the surface are small. In other words, as the frequency increases, the corresponding amplitude $G(f)$ decreases. Therefore, the terrain can be considered as a *band-limited signal* (i.e., it does not have any frequency components beyond specific frequency interval — $G(f) = 0$ for $f \in [-w,\, w]$); see Figure 4.10. The following sections briefly illustrate the conceptual basis for the utilization of Fourier analysis to derive the *optimum sampling interval* of continuous signals (i.e., deriving discrete signals from continuous ones) as well as to resample the discrete signals to recover the original continuous signal.

Optimum Sampling

A DTM represents the continuous surface of the Earth using finite/discrete samples. The sampling process is considered optimum if and only if the original continuous signal can be derived from the discrete samples. In this section, we will conceptually demonstrate the necessary conditions for having an optimal sampling process by considering a one-dimensional profile without getting into mathematical details. These principles can be generalized to two-dimensional surfaces. A continuous signal $g(x)$ can be sampled at equal intervals Δx to produce the discrete signal g_k, (refer to Figure 4.11).

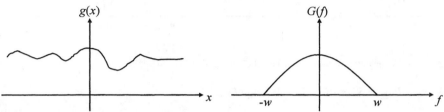

Figure 4.10 Fourier transform of a band-limited signal.

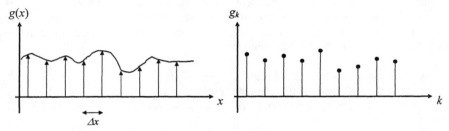

Figure 4.11 Sampling a continuous signal to produce a discrete one.

Using Fourier analysis, one can prove that the Fourier transform of the discrete signal g_k is nothing but a reproduction of the spectrum of the original continuous signal at the locations $\pm 1/\Delta x$, $\pm 2/\Delta x$, $\pm 3/\Delta x$, and so forth (Figure 4.12). A closer look at Figure 4.12 reveals that the spectra of the continuous signal will not overlap if and only if ($w < 1/2\Delta x$). Thus, the spectrum of the original signal, $G(f)$, can be recovered from the spectrum of the sampled one, $G_k(f)$, if and only if ($\Delta x < 1/2w$). In that case, the derived spectrum $G(f)$ can be incorporated into (4.23) to obtain the original continuous signal $g(x)$. The condition on the sampling interval (i.e., $\Delta x < 1/2w$) is known as the *Whittaker-Shannon sampling theorem*, which states that if the sampling interval Δx ensures that at least two samples are taken within every period of the highest frequency component of $g(x)$, then the original continuous signal can be recovered from the sampled data without any loss of information (Brown and Hwang, 1992). This sampling rate is known as the optimum sampling interval.

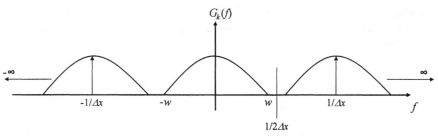

Figure 4.12 Spectrum of the discrete signal G_k in Figure 4.11.

Resampling in Spatial and Frequency Domains

In this section, we will be investigating the recovery of the continuous signal from the sampled one. As it has been established in the previous section, the original signal can be derived from the discrete one if and only if the sampling interval is smaller than one half of the reciprocal of the highest frequency component of the original signal. The resampling process in the frequency domain can be established according to the following steps:

1. Derive the spectrum of the sampled signal, $G_k(f)$.
2. Derive the spectrum of the original continuous signal $G(f)$ by multiplying the spectrum of the discrete signal, $G_k(f)$, with the *rectangular function* (Figure 4.13).
3. Use the inverse Fourier transform, (4.23), to derive the original continuous signal $g(x)$.

Since a multiplication in the frequency domain corresponds to a convolution in the spatial domain, the resampling/interpolation process can be carried out in the spatial domain as well. The rectangular function in the frequency domain, Figure 4.13, corresponds to the *sinc function* in Figure 4.14, which is defined over the domain $[-\infty \leq x \leq \infty]$. Thus, the value of the original function at an unsampled location, g_i, can be derived by convolving the discrete signal, g_k, with a sinc function, while centering the *sinc* function at the interpolation point (Figure 4.15). Note that the convolution in Figure 4.15 is nothing but a global weighted average of the sampled values, g_k, while the weights are dictated by the sinc function.

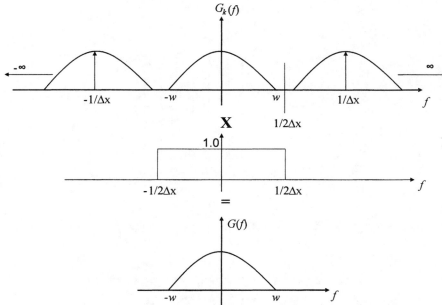

Figure 4.13 Deriving the spectrum of the continuous function, $G(f)$, from that associated with the sampled one, $G_k(f)$.

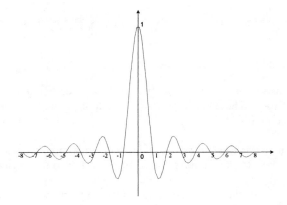

Figure 4.14 Sinc function $\{\sin(\pi x)/\pi x\}$.

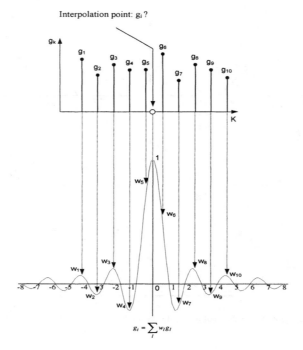

Figure 4.15 Resampling/interpolation through convolution of the discrete signal in the spatial domain with a sinc function.

Concluding Remarks

In this section, we presented the conceptual basis of using Fourier analysis for DTM interpolation. The most important characteristics of such a methodology can be summarized as follows:

1. Fourier interpolation can be only used when dealing with regularly distributed reference points since the Fourier transform is only defined for regularly spaced samples.
2. Interpolated elevations using spatial domain and frequency domain computations should be identical.
3. The interpolation process is exact. In other words, interpolated elevations at the locations of the reference points will be identical to the sampled elevations at these points.
4. Fourier interpolation is a global technique since it involves all the sample points either to derive the spectrum $G_k(f)$ in the frequency domain computation or to estimate the convolution in the spatial domain computation.

5. Frequency domain computations can be made efficient through the use of fast Fourier transform (FFT) techniques.

6. Since the amplitudes of the sinc function decreases as we move away from its center, the convolution computations in the spatial domain can be reduced by ignoring the samples whose weights are below a given threshold. It should be noted that this simplification would make the interpolation process a local rather than global one. This simplification will be discussed within the local interpolation techniques.

4.4.1.3 KRIGING

KRIGING is a geostatistical interpolation technique that estimates the elevation at the interpolation point as a weighted average of the observed elevations at the reference points (Clark, 1979, and Davis, 1986). This is similar to the TSA and Fourier analysis; refer to (4.12) and Figure 4.15, respectively. In the TSA, the weights are determined based on the XY-coordinates associated with the reference and interpolation points as well as the chosen deterministic function that describes the trend. Within Fourier analysis, the weights are dictated by the sinc function (Figures 4.14 and 4.15). However, the weights in the KRIGING interpolation are determined according to the distances between the interpolation and reference points as well as the stochastic properties of the surface. The stochastic properties of the surface, which describe the interaction between the elevations along the surface, are determined by analyzing the elevations at the reference points. This is advantageous when compared to the TSA, where the trend is assumed to follow a deterministic function that might or might not be compatible with the surface characteristics as defined by the reference points. However, this advantage comes at a price of increased computational complexity since a large set of simultaneous equations must be solved for every interpolation point estimated by KRIGING. Therefore, computer run time will be significantly longer if a surface is produced by KRIGING rather than by TSA. In addition, an extensive prior study of the data must be made to derive the stochastic properties of the surface in question. Chapter 5 presents a detailed discussion of the KRIGING interpolation methodology.

4.4.2 Local Interpolation Methods

Global interpolation methods impose external and global spatial structures on the interpolation procedure. Within these methods, local surface variations are considered to be random/unstructured noise. Intuitively, this might not be always the case as one expects that the interpolated Z value at any location to be similar to the measured Z values at nearby reference points. In the meantime, the interpolated Z value might be completely independent of the elevations at faraway reference points. Consequently, local interpolation methods only utilize the elevation information from the nearest reference points. In general, the implementation of a local interpolation technique involves the following activities:

1. Define a search area (neighborhood) around the point to be interpolated;
2. Identify the reference points within this neighborhood;
3. Choose a model to mathematically represent the elevation variation over this neighborhood;
4. Use the above model to evaluate the elevation at the point of interest.

Within the local interpolation procedure, one must address the kind of interpolation function (i.e., the type of the mathematical model) to use. In addition, a decision should be made regarding the size, shape, and orientation of the search neighborhood including the number and distribution of reference points to be incorporated. These issues will be investigated throughout the different interpolation functions, which can be classified as follows:

1. Interpolation from irregularly distributed reference points (e.g., TIN):
    *   Linear interpolation;
    *   Second exact fit surface interpolation;
    *   Quintic interpolation.

2. Interpolation from grid/irregular data:
    *   Nearest neighbor;
    *   Linear interpolation;
    *   Bilinear interpolation;
    *   Cubic convolution;
    *   Inverse distance weighting (IDW).

4.4.2.1 Interpolation from Irregularly Distributed Reference Points

The following interpolation methods will estimate the elevation at a nonsampled point using irregularly distributed reference points, such as those defined by a TIN. Note that the following techniques are exact interpolators.

Linear Interpolation

The linear interpolation method approximates the terrain by a continuous faceted surface formed by the triangles of a TIN. The surface value to be estimated is calculated based on the Z values at the nodes of the triangle within which the interpolation point lies. Thus, the interpolated elevation is not affected by the surface behavior of adjacent triangles. The basic assumption in linear interpolation is that each triangle defines a planar surface patch. The interpolation process can be carried out through vector algebra or plane fitting as explained in the following.

Linear Interpolation Using Vector Algebra

Let us assume that the vectors $\vec{a}$, $\vec{b}$, *and* $\vec{c}$ define the nodes of the triangle under consideration (A, B, and C, respectively). These vectors will be used to estimate the elevation Z_P at the given location (X_P, Y_P) of the interpolation point P in Figure 4.16. The normal vector to the triangle plane $\vec{n}$ can be derived through the cross-product in (4.24); refer to Figure 4.16 (Smith, 1998).

$$\vec{n} = (\vec{c} - \vec{a}) \times (\vec{b} - \vec{a}) \tag{4.24}$$

Since the vector ($\vec{p} - \vec{a}$) is contained within the triangle plane, it should be perpendicular to the normal vector $\vec{n}$ and satisfy the constraint in (4.25).

$$\vec{n} \bullet (\vec{p} - \vec{a}) = (\vec{c} - \vec{a}) \times (\vec{b} - \vec{a}) \bullet (\vec{p} - \vec{a}) = 0 \tag{4.25}$$

The expansion of the constraint in (4.25) is equivalent to the determinant in (4.26), with the elevation at the interpolation point (Z_P) as the only unknown.

$$\begin{vmatrix} X_A & Y_A & Z_A & 1 \\ X_B & Y_B & Z_B & 1 \\ X_C & Y_C & Z_C & 1 \\ X_p & Y_p & Z_p & 1 \end{vmatrix} = 0 \tag{4.26}$$

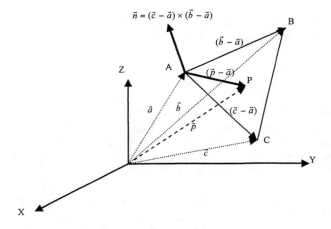

Figure 4.16 Vector representation of a plane through three points.

Linear Interpolation Using Plane Fitting
As mentioned earlier, linear interpolation assumes that the triangle including the interpolation point is represented by a planar surface (first-order polynomial) as in (4.27).

$$Z(X, Y) = a_0 + a_1 X + a_2 Y \qquad (4.27)$$

The nodes of the triangle (*A*, *B*, and *C*) should satisfy the equation of the defined plane. This will lead to the linear system in (4.28), which can be used to solve for the coefficients (a_0, a_1, a_2) describing the plane through the triangle under consideration.

$$\begin{bmatrix} Z_A \\ Z_B \\ Z_C \end{bmatrix} = \begin{bmatrix} 1 & X_A & Y_A \\ 1 & X_B & Y_B \\ 1 & X_C & Y_C \end{bmatrix} \begin{bmatrix} a_0 \\ a_1 \\ a_2 \end{bmatrix} \qquad (4.28)$$

Finally, the elevation at the interpolation point, Z_P, can be estimated by substituting its *XY*-coordinates in the plane equation, (4.29).

$$Z_P = \begin{bmatrix} 1 & X_P & Y_P \end{bmatrix} \begin{bmatrix} a_0 \\ a_1 \\ a_2 \end{bmatrix} \qquad (4.29)$$

Note that linear interpolation is an exact interpolator, which produces a continuous but not smooth surface. The surface normal will abruptly change as we move from one triangle to the next.

Second Exact Fit Surface Interpolation

The assumption of the linear interpolation that the triangles are represented by tilted flat facets results in a very crude surface. A better approximation can be achieved using curved or bent triangular surfaces, especially if these can be made to join smoothly across the edges of the triangles. The second exact fit surface is an interpolation methodology that aims at satisfying this objective. It starts by finding the three triangular neighbors closest to the triangle including the interpolation point (Figure 4.17). Then, a second-order polynomial trend is fitted through the six nodes of the identified triangles. The triangle nodes should satisfy the equation of the chosen trend resulting in the linear system in (4.30), which can be used to solve for the polynomial coefficients $(a_0, a_1, a_2, a_3, a_4, a_5)$. Finally, the estimated coefficients and the *XY*-coordinates of the interpolation point (X_P, Y_P) can be used to evaluate the elevation (Z_P) at the interpolation point.

$$
\begin{bmatrix} Z_1 \\ Z_2 \\ Z_3 \\ Z_4 \\ Z_5 \\ Z_6 \end{bmatrix} = \begin{bmatrix} 1 & X_1 & Y_1 & X_1^2 & X_1Y_1 & Y_1^2 \\ 1 & X_2 & Y_2 & X_2^2 & X_2Y_2 & Y_2^2 \\ 1 & X_3 & Y_3 & X_3^2 & X_3Y_3 & Y_3^2 \\ 1 & X_4 & Y_4 & X_4^2 & X_4Y_4 & Y_4^2 \\ 1 & X_5 & Y_5 & X_5^2 & X_5Y_5 & Y_5^2 \\ 1 & X_6 & Y_6 & X_6^2 & X_6Y_6 & Y_6^2 \end{bmatrix} \begin{bmatrix} a_0 \\ a_1 \\ a_2 \\ a_3 \\ a_4 \\ a_5 \end{bmatrix}
\tag{4.30}
$$

Note that the fitted surface will exactly pass through the triangle nodes, yielding an exact interpolator. Contour maps derived from this method will be curved rather than straight lines as in the linear interpolation method. Even though adjacent triangles are fitted using common points, their trend surfaces will not smoothly coincide along lines of overlap. This means there might be abrupt changes in direction when crossing from one triangle to the next (i.e., we will end up with a nonsmooth surface).

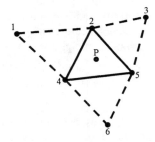

Figure 4.17 Involved triangles in a second exact fit surface interpolation.

Quintic Interpolation

Similar to the linear and second exact fit surface interpolation methods, quintic interpolation considers the surface model to be continuous. In addition, it considers the surface model to be smooth. Therefore, the normal to the surface varies continuously within each triangle and there will be no abrupt changes in the normal as one crosses an edge between neighboring triangles. The smoothness characteristic is accomplished by considering the geometry of the neighboring triangles when interpolating for the *Z* value of a point in a triangle. Quintic interpolation employs a fifth-degree trend, (4.31).

$$Z(X,Y) = \sum_{i=0}^{5} \sum_{j=0}^{5-i} a_{ij} X^i Y^j \tag{4.31}$$

The trend surface in (4.31) involves 21 coefficients (a_0, a_1, ..., a_{20}). The values of the fifth-order function, its first-order, and second-order partial derivatives (Z, $\partial Z / \partial X$, $\partial Z / \partial Y$, $\partial^2 Z / \partial X^2$, $\partial^2 Z / \partial Y^2$, $\partial^2 Z / \partial X \partial Y$) are estimated at each node of the triangle using the elevation values at that node as well as neighboring triangles, yielding 18 equations. Another three equations are established by considering the surface to be smooth and continuous in the direction perpendicular to the three triangle edges. Thus, we will end up with 21 linear equations involving 21 unknowns. Finally, the estimated coefficients and the XY-coordinates of the interpolation point (X_P, Y_P) can be used to evaluate the elevation (Z_P) at the interpolation point. For more technical details regarding quintic interpolation, interested readers can refer to Akima, 1978.

4.4.2.2 Interpolation from Regularly and/or Irregularly Distributed Reference Points

The following sections will elaborate on local and deterministic interpolation methods that can deal with regularly and irregularly distributed reference points. The discussed techniques include nearest neighbor, linear, bilinear, cubic convolution, and inverse distance weighting interpolation methods. These techniques share the same property of being an exact interpolator.

Nearest Neighbor

The nearest neighbor technique assigns the elevation associated with the closest reference point to the interpolation point under consideration. For example, Figure 4.18 illustrates the concept of the nearest neighbor interpolation when dealing with regularly distributed reference points. Note that there is no averaging/smoothing of the elevations at neighboring reference points. In other words, interpolated elevations will assume elevation values that are identical to those associated with the reference points. The main advantage of the nearest neighbor is its computational efficiency. On the negative side, it produces discontinuous surfaces. In addition to DTM interpolation, nearest neighbor is popular in image processing applications, especially those dealing with classification activities. The main advantage of the nearest neighbor is maintaining the original gray values in the original imagery (Jahne, 1997).

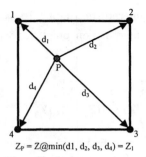

$$Z_P = Z@\min(d_1, d_2, d_3, d_4) = Z_1$$

Figure 4.18 Principle of nearest neighbor interpolation.

Linear Interpolation

Similar to the TIN linear interpolation, we use the closest three reference points to define a planar surface. The fitted plane is then used to estimate the elevation at the interpolation point in question. Therefore, the whole interpolation domain is subdivided into triangles whose corners are at the reference points. Figure 4.19 illustrates the conceptual basis of linear interpolation from raster data with ΔX and ΔY as the spacing between the grid elements along the X and Y directions, respectively. In such a case, the interpolation can proceed as follows:

1. Search for the grid element where the interpolation point (X, Y) lies. In other words, determine the vertices (X_1, Y_1, Z_1), (X_2, Y_2, Z_2), (X_3, Y_3, Z_3), and (X_4, Y_4, Z_4) of the grid element, including the interpolation point.
2. Normalize the XY-coordinates of the interpolation point according to (4.32). The normalized coordinates refer to the coordinate system whose origin is at the lower left reference point with the cell size scaled to unity [Figure 4.19(b)].
3. Determine the value of the variable δ as in (4.33). This variable indicates whether the interpolation point lies within the upper or lower triangle of the identified grid element in the first step.
4. Estimate the elevation at the interpolation point according to (4.34).

$$\bar{X} = \frac{X - X_1}{\Delta X}$$

$$\bar{Y} = \frac{Y - Y_1}{\Delta Y}$$

(4.32)

$$\delta = \begin{cases} 1 \mapsto \text{if } \overline{X} \le \overline{Y} & \text{The interpolation point lies within the } \Delta123 \\ 0 \mapsto \text{Otherwise} & \text{The interpolation point lies within the } \Delta134 \end{cases} \qquad (4.33)$$

$$Z(\overline{X}, \overline{Y}) = \delta\{Z_1 + (Z_3 - Z_2)\overline{X} + (Z_2 - Z_1)\overline{Y}\}$$
$$+ (1-\delta)\{Z_1 + (Z_4 - Z_1)\overline{X} + (Z_3 - Z_4)\overline{Y}\} \qquad (4.34)$$

Note that the linear interpolation produces a continuous surface that passes through the reference points (i.e., exact interpolator). However, the generated surface will not be a smooth one.

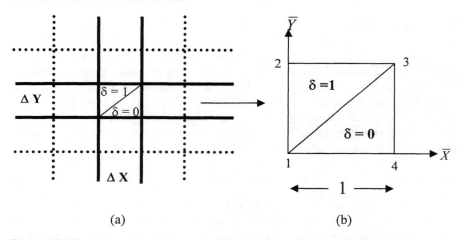

Figure 4.19 Linear interpolation within a grid: (a) original grid and (b) normalized grid.

Bilinear Interpolation

Instead of dividing the grid elements into triangles within linear interpolation, bilinear interpolation considers the whole grid element as the basic unit. A bilinear polynomial of the form in (4.35) is defined over the grid element including the interpolation point (Figure 4.20).

$$Z(X,Y) = a_0 + a_1 X + a_2 Y + a_3 XY \qquad (4.35)$$

Using the four reference points at the corners of the grid element including the interpolation point, we can define four equations. These equations can be used to evaluate the bilinear polynomial coefficients, which can be used to estimate the elevation at the interpolation point. Alternatively, the computational procedure of the bilinear interpolation can proceed as follows:

1. Search for the grid element where the interpolation point (X, Y) lies. In other words, determine the vertices (X_1, Y_1, Z_1), (X_2, Y_2, Z_2), (X_3, Y_3, Z_3), and (X_4, Y_4, Z_4) of the grid element, including the interpolation point.
2. Normalize the XY-coordinates of the interpolation point according to (4.36). The normalized coordinates refer to the coordinate system whose origin is at the lower left reference point with the cell size scaled to unity (Figure 4.20).
3. Estimate the elevation at the interpolation point according to (4.37).

$$\bar{X} = \frac{X - X_1}{\Delta X}$$
$$\bar{Y} = \frac{Y - Y_1}{\Delta Y} \tag{4.36}$$

$$Z(\bar{X}, \bar{Y}) = Z_1 + (Z_4 - Z_1)\bar{X} + (Z_2 - Z_1)\bar{Y} + (Z_1 - Z_2 + Z_3 - Z_4)\overline{XY} \tag{4.37}$$

Bilinear interpolation will produce a continuous and nonsmooth surface that passes through the reference points (i.e., exact interpolator). Note that the bilinear interpolation is conceptually similar to the discussed convolution in Section 4.4.1.2 (refer to Figure 4.15). The only difference is that the bilinear interpolation approximates the two-dimensional sinc function, Figure 4.14, with a two-dimensional triangular function (a one-dimensional cross section is shown in Figure 4.21).

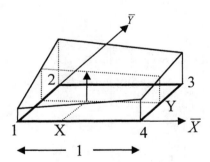

Figure 4.20 Bilinear interpolation.

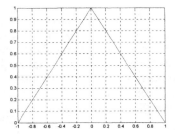

Figure 4.21 One-dimensional bilinear convolution kernel.

Cubic Convolution

Cubic convolution calculates the output Z value as a weighted average of the elevations at the closest 16 reference points. The weighting function is defined by a two-dimensional cubic convolution kernel (one-dimensional cross section is shown in Figure 4.22). Using a smooth curve and a larger neighborhood gives the cubic convolution a tendency to smooth the data. Note that the cubic convolution is more similar to the discussed convolution in Section 4.4.1.2 (refer to Figure 4.15) than the bilinear interpolation. The closer similarity is attributed to the fact that the cubic convolution kernel is a better approximation of the *sinc* function than the triangular function; compare Figures 4.14, 4.21, and 4.22. Similar to the linear and bilinear interpolations, cubic convolution is an exact interpolator that produces a continuous surface, which might not be smooth.

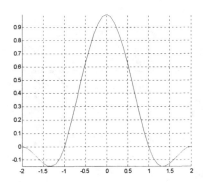

Figure 4.22 One-dimensional cubic convolution kernel.

Inverse Distance Weighting

Similar to the above interpolation techniques, inverse distance weighting estimates the Z value at a point as a weighted average of the elevations at nearby reference points. The weights are inversely proportional to the distance between the interpolation point and the reference point in question. In other words, nearby reference points will have higher weight/influence than more distant data. The conceptual basis behind this technique is that nearby points have similar elevation values, while the elevations at faraway points are almost independent. The weighting function and averaging process can be mathematically described by (4.38):

$$Z(X, Y) = \sum_{i=1}^{n} \left[\frac{Z_i}{d_i^p} \right] \bigg/ \sum_{i=1}^{n} \left[\frac{1}{d_i^p} \right] ; \text{ or}$$

$$Z(X, Y) = \sum \lambda_i \bullet Z_i \xrightarrow{\text{With}} \sum \lambda_i = 1$$

(4.38)

$Z(X, Y)$, in the above equation, is the estimated Z value at the location (X, Y) while Z_i is the elevation value at the ith reference point located at (X_i, Y_i), (d_i) is the distance between the reference and the interpolation points, and (p) is the power to which the distance is raised. Depending on the site conditions, the distance may be weighted in different ways. If the (p) value is selected to be 1, then we will end up with a simple linear interpolation between points. Practical experience has showed that selecting $(p = 2)$ produces better results. In this case, close points are heavily weighted, and more distant points are lightly weighted. To avoid numerical problems resulting from having the interpolation point coinciding with a reference point (i.e., $d = 0$), one should incorporate the following condition in the interpolation; if $d_i = 0$, then $Z(X, Y) = Z_i$. Note that the inverse distance weighting can be used for interpolation while considering surface constraints. For example, during the interpolation, a weight of zero should be selected for the reference points if these points and the interpolation point in question lie on opposite sides of a breakline. In other words, reference points on the opposite side of a breakline are not permitted to contribute toward the elevation at the interpolation point.

4.4.3 Data Gridding

Gridding is the process of estimating the surface elevations at a set of locations that are arranged in a regular pattern from irregularly distributed reference points. The locations where the elevations will be estimated are referred to as *grid points* or *grid nodes*. The grid nodes are usually arranged in a square pattern (i.e., the distance between the nodes in one direction is the same as the distance between them in the perpendicular direction). The area enclosed by four grid nodes is known as a *grid cell*. If a large size is chosen for the grid cells, the resulting surface will have low resolution and a coarse appearance, but can be computed quickly. Conversely, if the grid cells are small in size, the resulting surface will have a finer appearance, but will be more time-consuming and expensive to produce.

The gridding process involves three essential steps. First, the reference points must be sorted according to their *XY*-coordinates. Second, considering a specific grid node, one must search for the neighboring reference points. Third, one needs to estimate the elevation value of that grid node by some mathematical function, as described in the previous sections, using the elevations at the neighboring reference points. The implemented sorting technique greatly affects the speed of the gridding process. However, it has no effect on the accuracy/shape of the estimated surface. The shape and the quality of the final surface depend on the mathematical function as well as the defined neighborhood. The following sections present some gridding and search techniques.

4.4.3.1 Moving Average

The most obvious interpolation function that can be used to estimate the elevation at a specific grid node is to simply calculate an average of the elevations at nearby reference points. The averaging function should weigh the closest reference points more heavily than distant points. The resulting surface will have certain characteristics. For example, the highest and lowest areas on the surface will contain the reference points, and the grid nodes will have intermediate values, since an average cannot be outside the range of the numbers from which it is derived.

To illustrate the concept of the gridding process, let us consider Figure 4.23(a), which shows the elevations at a set of reference points. The observations may be identified and sorted by numbering them sequentially as they are encountered by proceeding from left to right and top to bottom. Figure 4.23(b) depicts a regular grid of nodes that has been superimposed on the area covered by the reference points. To estimate the elevation of a given grid node, the nearest (n) reference points must be found and the distances between them and the node are calculated [Figure 4.23(c)]. The search procedure may be simple or elaborate. Having found the distances to the nearest reference points, the grid point elevation Z is estimated. The completed grid with interpolated elevations is shown in Figure 4.23(d).

This gridding methodology is sometimes referred to as *moving average*, since each node in the grid is estimated as the average of the elevations at the reference points within a neighborhood, which is *moved* from one grid node to the next. In effect, the elevations at the reference points are projected horizontally to the location of the grid node, where they are weighted and averaged.

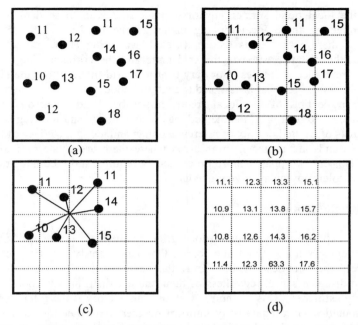

Figure 4.23 Computational procedure for grid interpolation using moving average: (a) original reference points, (b) selected grid, (c) neighboring reference points, and (d) interpolated grid.

4.4.3.2 Linear Projection Gridding

The linear projection gridding is a two-part procedure, in which a weighted average of projected local trends from neighboring reference points around each grid node are used to estimate the elevation value at that node. The first part estimates the local trend surfaces at the locations of the reference points. Then, the computed local trends are used to estimate the elevation at the grid nodes.

To estimate the local trend at a specific reference point, the closest (n) reference points are found and each is weighted inversely to its distance from the reference point in question. Then, a linear trend surface is fitted to these weighted elevations. The constant term of the trend equation is adjusted to make the fitted plane exactly pass through the reference point. If at least five reference points cannot be found around the reference point or if the simultaneous equations for the fitted plane cannot be solved, the coefficients of a global linear trend are used to estimate the local trend. The trend coefficients are saved for each reference point.

The second part of the algorithm estimates the elevation value of the surface at the grid nodes. A search procedure finds the nearest (n) reference points around the node to be estimated. The XY-coordinates of the grid node are substituted into each of the local trend surface equations associated with these data points; in effect, projecting these local dipping planes to the location of the node. An average of these estimates is then calculated, weighting each estimate by the inverse of the distance between the grid node and the corresponding reference point. If a reference point lies at or very near to a grid intersection, the elevation value of the reference point is assigned to that grid node.

The projection of the local trends might be disadvantageous in some circumstances. For example, the method may tend to create spurious highs or lows both in areas of limited density of reference points and along the edges of the map if surface trends are projected from areas where there are clusters of reference points. Also, the two-phase algorithm obviously requires more computation time than the simpler moving average procedure.

4.4.3.3 Search Algorithms

One critical difference between various gridding methods is the way in which nearest neighbors are defined and found. One can identify a variety of search procedures. The simplest method finds the nearest (n) reference points, in a Euclidean distance sense, regardless of their angular distribution around the grid node being estimated. This method is fast and satisfactory if the reference points are distributed in a comparatively uniform manner. An objection to the simple nearest neighbor search is that all the nearest points may lie in a narrow wedge on one side of the grid node. In this case, the interpolated elevation at the node is essentially unconstrained, except in one direction. This may be avoided by restricting the search in some way that ensures the reference points are equitably distributed around the grid node in question.

The simplest method that introduces some restrictions/constraints is a *quadrant search* [Figure 4.24(a)]. Some minimum number of reference points must be selected from each of the four quadrants around the grid node being calculated. An elaboration on the quadrant search is an *octant search* [Figure 4.24(b)], which introduces a further constraint on the radial distribution of the points used in the interpolation process. A specified number of control points must be found in each of the 45° segments surrounding the grid node being estimated. The constrained search procedures require finding and testing more neighboring reference points than in a simple search, thus increasing the processing time.

The above search procedures are based on a fixed number of reference points collected at variable distances from the grid node. An alternative search methodology is to identify all the reference points within a radius (r) relative to the grid node being estimated. This search algorithm uses a variable number of points found within a fixed distance from the node under consideration. The advantage of a distance-based search over constrained search algorithms is that any constraints on the search for the nearest reference points, such as a quadrant or octant requirement, will obviously expand the size of the neighborhood around the grid node being estimated. Therefore, some nearby reference points are likely to be passed over in favor of more distant points in order to satisfy the requirement that only a few points may be taken from a single sector. Since remote reference points are not related to the elevation at the interpolation point, the resulting surface from constrained search procedures will be poorer when compared to that generated from a simple distance-based search algorithm.

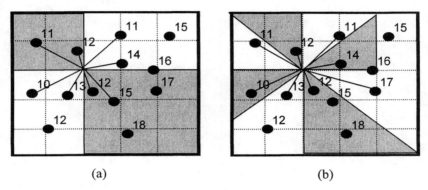

(a)   (b)

Figure 4.24 (a) Quadrant and (b) octant constrained search procedures.

References

Akima, H., "A method of bivariate interpolation and smooth surface fitting for irregularly distributed data points," *Transactions on Mathematical Software*, 4(2): 148-159, 1978.

Brown, R., and Hwang, P., *Introduction to Random Signals and Applied Kalman Filtering*, New York: John Wiley & Sons, 1992.

Clark, I., *Practical Geostatistics*, London: Applied Science Publishers, 1979.

Davis, J., *Statistics and Data Analysis in Geology*, New York: John Wiley & Sons, 1986.

Jahne, B., *Digital Image Processing: Concepts, Algorithms, and Scientific Applications*, Berlin: Springer-Verlag, 1997.

Koch, K., *Parameter Estimation and Hypothesis Testing in Linear Models*, Berlin: Springer-Verlag, 1987.

Oppenheim, A., and Schafer, R., *Discrete Signal Processing*, Upper Saddle River, NJ: Prentice-Hall, 1989.

Rice, J., *Mathematical Statistics and Data Analysis*, Monterey, CA: Wadsworth & Brooks, 1987.

Smith, L., *Linear Algebra*, Berlin: Springer-Verlag, 1998.

Chapter 5

KRIGING: A Closer Look

5.1 INTRODUCTION

So far, none of the interpolation methods discussed in the previous chapter can provide a direct estimate of the quality of the interpolation made in terms of a variance for the interpolated elevation at a certain location. In all methods, the only way to determine the goodness of the interpolation would be to compute estimates for a set of extra checkpoints that had not been used in the original interpolation (this is rarely done in practice).

A further objection of all the previous interpolation methods is that there is no a priori method of determining whether the best values have been chosen for the weighting parameters or if the size of the search neighborhood is appropriate. Moreover, no method studied so far provides sensible information on (Davis, 1986):

1. The number of points needed to compute the local coverage;
2. The size, orientation, and shape of the neighborhood from which those points are drawn;
3. Whether there are better ways to estimate the interpolation weights than as a simple deterministic function of distance;
4. The errors associated with the interpolated values.

KRIGING is a geostatistical interpolation technique that addresses the above-mentioned questions and concerns. Conceptually, KRIGING is similar to previously discussed interpolation techniques in the sense that an estimate of the elevation at the interpolation point is determined as a weighted average of the observed elevations at the reference points. The main difference between KRIGING and other interpolation methods is that the weights are determined according to the distances between the interpolation and reference points as well as the stochastic properties of the surface. The discussion starts by establishing some terminology, which will be used throughout this chapter. Next, we proceed by deriving the KRIGING formulas using least squares adjustment principles. Then, KRIGING interpolation is established by determining the *best linear estimate* (BLE) or the *best linear unbiased estimate* (BLUE) of the elevation at the interpolation points. Empirical determination of the stochastic properties of the surface, using variance-covariance or semivariance analysis, is discussed afterwards. Finally, the concept and implementation of the universal KRIGING are briefly outlined.

5.2 TERMINOLOGY

Before getting into the technical details of KRIGING, we start by listing some symbols, associated terminology, and principles, which are pertinent to least squares adjustment and DTM interpolation. For more technical details, interested readers can refer to Koch (1987), Mikhail (1976), Mikhail and Gracie (1981), and Wolf and Ghilani (1997).

Observation vector (y): For DTM interpolation, the observation vector will be used to represent the sampled elevations at the reference points, which will be denoted by Z_r.

Unknown vector (x): For DTM interpolation, the unknown vector will be used to represent the sought elevations at the interpolation points, which will be denoted by Z_p.

Design matrix (A): This matrix will be used to denote the linear mathematical relationship between the observation and unknown vectors.

Error vector (e): This vector will be used to denote the random noise contaminating the observation vector (*y*). The error vector is mathematically defined by the difference between the observed quantities and the corresponding true values.

Expectation (E): The expectation is used to denote the mean of the quantity in question. The expectation of a truly random error vector is zero (i.e., $E\{e\} = 0$).

Dispersion matrix (D): The dispersion matrix of a given vector describes the quality of its individual components. The dispersion matrix is square with the diagonal elements defining the variances while the off-diagonal elements define the covariance between the individual elements of the vector in question. The dispersion matrix is sometimes referred to as the *variance-covariance matrix*. In general, the dispersion matrix can be mathematically defined as follows:

$$[D\{y_{n\times1}\}]_{n\times n} = E\{(y - E\{y\})_{n\times1} \ (y - E\{y\})_{1\times n}^{T}\} \tag{5.1}$$

The dispersion of the error vector (e) is usually expressed as the product of a scalar quantity (σ_o^2) and square matrix (P^{-1}) (i.e., $D\{e\} = \sigma_o^2 \ P^{-1}$).

Covariance matrix (C): The covariance matrix describes the covariance between elements of two vectors. The covariance matrix can be mathematically defined as in (5.2). If the two vectors are not correlated, their covariance matrix will be zero.

$$[C\{x_{m\times1}, \ y_{n\times1}\}]_{m\times n} = E\{(x - E\{x\})_{m\times1} \ (y - E\{y\})_{1\times n}^{T}\} \tag{5.2}$$

Law of error propagation: If the vectors x and y are related to each other as in (5.3), then the dispersion $D\{y\}$ can be derived through the law of error propagation as shown in (5.4):

$$y_{n\times1} = B_{n\times m} \ x_{m\times1} \tag{5.3}$$

$$D\{y\}_{n\times n} = B_{n\times m} \ D\{x\}_{m\times m} \ B_{m\times n}^{T} \tag{5.4}$$

The law of error propagation can be used to derive the dispersion of the summation of two quantities as follows:

$$D\{x + y\} = D\{\begin{bmatrix} 1 & 1 \end{bmatrix}\begin{bmatrix} x \\ y \end{bmatrix}\} = \begin{bmatrix} 1 & 1 \end{bmatrix}\begin{bmatrix} D\{x\} & C\{x,y\} \\ C\{y,x\} & D\{y\} \end{bmatrix}\begin{bmatrix} 1 \\ 1 \end{bmatrix}$$

$$D\{x + y\} = D\{x\} + D\{y\} + 2 \ C\{x, y\} \tag{5.5}$$

If x and y are not correlated, then the dispersion matrix would reduce to the one in (5.6):

$$D\{x + y\} = D\{x\} + D\{y\} \tag{5.6}$$

5.3 KRIGING FROM THE LEAST SQUARES ADJUSTMENT PRINCIPLES

The Gauss-Markov model (observations equations) is represented by the following equation (Koch, 1987):

$$y_{n \times 1} = A_{n \times m} \, x_{m \times 1} + e_{n \times 1} \quad e \sim (0, \sigma_o^2 \, P^{-1}) \tag{5.7}$$

For the surface interpolation problem, (y) represents the measured elevations at the reference points with a variance-covariance matrix $D\{y\} = \sigma_o^2 \, P^{-1}$. On the other hand, (x) represents the unknown elevations at the interpolation points. The estimate for the unknown parameters that minimizes the weighted squared sum of the error vector ($e^T P e$) is defined as follows:

$$\hat{x}_{m \times 1} = \left(A^T P A \right)^{-1}_{m \times m} (A^T P y)_{m \times 1}$$
$$\hat{x}_{m \times 1} = N^{-1}_{m \times m} \, C_{m \times 1} \tag{5.8}$$

Now, let us investigate the covariance matrix of the observed and estimated vectors $(y_{n \times 1}, \hat{x}_{m \times 1})$. The covariance matrix can be derived through the law of error propagation as follows (Koch, 1987):

$$\begin{Bmatrix} y \\ \hat{x} \end{Bmatrix} = \begin{bmatrix} I_n \\ N^{-1} A^T P \end{bmatrix} y \tag{5.9a}$$

$$D\begin{Bmatrix} y \\ \hat{x} \end{Bmatrix} = \begin{bmatrix} D\{y\} & C\{y, \hat{x}\} \\ C\{\hat{x}, y\} & D\{\hat{x}\} \end{bmatrix} = \begin{bmatrix} I_n \\ N^{-1} A^T P \end{bmatrix} D\{y\} \begin{bmatrix} I & PAN^{-1} \end{bmatrix} \tag{5.9b}$$

$$D\begin{Bmatrix} y \\ \hat{x} \end{Bmatrix} = \begin{bmatrix} D\{y\} & C\{y, \hat{x}\} \\ C\{\hat{x}, y\} & D\{\hat{x}\} \end{bmatrix} = \begin{bmatrix} \sigma_o^2 P^{-1} & \sigma_o^2 A N^{-1} \\ \sigma_o^2 N^{-1} A^T & \sigma_o^2 N^{-1} \end{bmatrix} \tag{5.9c}$$

where I_n is an "$n \times n$" identity matrix. From the previous equation, the covariance between the estimated unknowns and the observed quantities can be described by $C(\hat{x}, y) = \sigma_o^2 \, N^{-1} \, A^T$. Using this covariance, we can rewrite the least squares estimate of the unknowns, (5.8), as follows:

$$\hat{x}_{m\times1} = N^{-1}A^T Py$$

$$\hat{x}_{m\times1} = (\sigma_o^2 N^{-1}A^T)(\sigma_o^{-2} P)y \qquad (5.10)$$

$$\hat{x}_{m\times1} = C\{\hat{x}, y\}_{m\times n} [D\{y\}]_{n\times n}^{-1} y_{n\times1}$$

Thus, the estimated unknown vector is the product of the covariance matrix between the unknown and observation vectors (which is a measure of the interaction between the unknown and observed quantities), the inverse of the variance-covariance matrix associated with the observation vector (which is a measure of the interaction between the observed quantities), and the observation vector. In other words, one can visualize the least squares adjustment solution as a weighted average of the observation vector, where the weights depend on the interaction between the unknowns and the observations as well as the interaction between the observations. This is the conceptual basis behind KRIGING.

5.4 KRIGING AS THE BEST LINEAR ESTIMATES (BLE)

Here, we would like to derive an estimate of the unknown parameters (the elevations at the interpolation points) as the best linear estimate (BLE). In this case, linear estimate indicates that the unknowns are estimated as the weighted sum of the observed quantities (i.e., $\hat{x}_{m\times1} = w_{m\times n}^T y_{n\times1}$). On the other hand, best indicates that the errors associated with the estimated unknowns have the minimum variance within the class of linear estimates. Based on the above criteria, one can determine the weight matrix ($w_{m\times n}^T$). Before deriving the weight matrix, let us assume that the observed vector is the elevations at the reference points (i.e., $y_{n\times1} = z_{r_{n\times1}}$). Also, let us assume that we are interested in the elevation at a single interpolation point (i.e., $x_{1\times1} = z_{P_{1\times1}}$). Therefore, the interpolated height as a weighted sum of the elevations at the reference points can be represented by:

$$\hat{z}_P = w_{1\times n}^T z_{r_{n\times1}} \qquad (5.11)$$

Also, the estimation error is given as:

$$e_P = z_P - \hat{z}_P \qquad (5.12)$$

where:

z_P is the true elevation at the interpolation point;

$\hat{z}_p$ is the estimated elevation at the interpolation point;

e_p is the estimation error at the interpolation point;

$z_{r_{m \times 1}}$ is the vector of observed elevations at the reference points.

The variance-covariance matrix for the estimation error at the interpolation point can be computed through the law of error propagation as follows:

$$e_p = z_p - \hat{z}_p = \begin{bmatrix} 1 & -w^T \end{bmatrix} \begin{bmatrix} z_p \\ z_r \end{bmatrix} \tag{5.13a}$$

$$D\{e_p\} = \begin{bmatrix} 1 & -w^T \end{bmatrix} \begin{bmatrix} D\{z_p\}_{1 \times 1} & C\{z_p, z_r\}_{1 \times n} \\ C\{z_r, z_p\}_{n \times 1} & D\{z_r\}_{n \times n} \end{bmatrix} \begin{bmatrix} 1 \\ -w \end{bmatrix} \tag{5.13b}$$

$$D\{e_p\} = D\{z_p\} - 2 w^T C\{z_r, z_p\} + w^T D\{z_r\} w \tag{5.13c}$$

where $D\{z_p\}$ is the variance of the elevation at the interpolation point, $C\{z_r, z_p\}$ is the covariance between the elevations at the reference and the interpolation points, and $D\{z_r\}$ is the variance-covariance matrix of the elevations at the reference points. Note that these variance-covariance matrices represent the stochastic properties, which depend on the nature of the surface in question. In other words, they describe the interaction between the elevations at various points along the surface. More specifically, they do not describe the nature of the random errors contaminating observed quantities.

To achieve the best estimate, one should choose the weighting vector that minimizes the above variance-covariance matrix. In other words, the weighting vector should satisfy the target function $\varphi(w)$ in (5.14a).

$$\varphi(w) = D\{z_p\} - 2 w^T C\{z_r, z_p\} + w^T D\{z_r\} w = \min(w) \tag{5.14a}$$

$$\frac{\partial \varphi}{\partial w} = -2 C\{z_r, z_p\} + 2 D\{z_r\} w = 0 \tag{5.14b}$$

$$w = \left[D\{z_r\} \right]^{-1} C\{z_r, z_p\} = \left[D\{z_r\} \right]^{-1} C\{z_r, z_p\} \equiv \left[D\{y\} \right]^{-1}_{n \times n} c\{y, \hat{x}\}_{n \times 1} \tag{5.14c}$$

Finally, the estimated height at the interpolation point can be derived as:

$$\hat{z}_p = w^T z_r = C\{z_P, z_r\} [D\{z_r\}]^{-1} z_r \tag{5.15}$$

Comparing (5.8), (5.10), and (5.15), one can see that the least squares solution to the Gauss Markov model is nothing but the best linear estimate, which is commonly known as the BLE. As mentioned earlier, the BLE is the product of the covariance matrix between the unknown and observation vectors (which is a measure of the interaction between the elevations at the interpolation and reference points), the inverse of the variance-covariance matrix associated with the observation vector (which is a measure of the interaction between the elevations at the reference points), and the observation vector (which is the observed elevations at the reference points).

An important advantage of the KRIGING is that we can derive a quantitative measure describing the quality of the estimated height at the interpolation point (which could not be derived for the various interpolation methods discussed in the previous chapter). By substituting (5.14c) into (5.13c), we can get the variance-covariance matrix for the estimation error as follows:

$$D\{e_p\} = D\{z_p\}_{1\times 1} - \left(w^T C\{z_r, z_p\}\right)_{1\times 1} \tag{5.16}$$

Finally, using the law of error propagation, we can compute the variance of the estimated height at the interpolation point as follows:

$$\hat{z}_p = w^T z_r$$
$$D\{\hat{z}_p\} = w^T D\{z_r\} w = w^T C\{z_r, z_p\} \tag{5.17}$$

By comparing (5.16) and (5.17), we can see that

$$D\{e_p\} = D\{z_p\} - D\{\hat{z}_p\} \tag{5.18}$$

A closer look at (5.12) and (5.18) reveals the fact that the interpolated elevation and the estimation error are not correlated, $\left(\text{i.e., } C\{e_p, \hat{z}_p\} = 0.0\right)$. This should come as no surprise since we should not expect any correlation between the estimated elevation and the purely random error vector. The above formulation is known within the literature as *simple KRIGING*. The question now is how can we obtain the dispersion and covariance matrices needed for the interpolation process, namely $D\{z_r\}$ and $C\{z_r, z_p\}$? The following section deals with empirical determination of such matrices from the measured elevations at the reference points.

5.5 EMPIRICAL DETERMINATION OF THE VARIANCE-COVARIANCE FUNCTION

Before getting into the details of determining the variance-covariance function, which can be used to derive $D\{z_r\}$ and $C\{z_r, z_p\}$, let us try to answer the following questions:

1. What is the independent variable of the function expressing the interaction (variance-covariance) between the elevations at various points along the surface under consideration?
2. What is the expected shape of the interaction (variance-covariance) function between the elevations at various points along the surface?

To answer these questions, one should note that the variance-covariance function represents the degree of similarity between the elevations at various points along the surface. It can be argued that the degree of similarity between the elevations depends on the horizontal distance between the points in question. One would expect that neighboring points would have similar elevations. On the other hand, faraway points might have significantly different elevations. Therefore, horizontal distance between the points is the independent variable of the variance-covariance function. Also, the variance-covariance function should have large value for small distances since neighboring points should have similar elevations. The value of the function should then decrease as the distance between the involved points increases (see Figure 5.1).

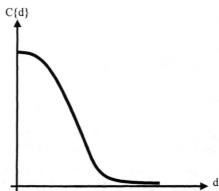

Figure 5.1 Expected shape of the variance-covariance function relating the elevations at points that are d-distance apart from each other.

To obtain the variance-covariance function, let us look into the nature of the elevations along the surface. We cannot deal with the elevations either as *deterministic* or *random variables*. If the elevations were deterministic, they could have been represented by an analytical function that describes the change in the elevations along the surface. On the other hand, they are not random, since if we have the elevation observed at one location, we have a reasonable idea about the elevation at neighboring points since the surface tends to be relatively smooth. As a result, the elevation can be treated as a *regionalized variable/random process*. Random processes might be *stationary* and *ergodic* (Brown and Hwang, 1992, and Gelb, 1979).

The stochastic properties of a stationary random process are invariant with respect to spatial or temporal shifts. For example, the expectation (mean) of a stationary random process will have the following property:

$$E\{X\} = E\{X+\tau\} \tag{5.19}$$

Where X is the random process in question and τ is a given spatial/temporal offset. Therefore, elevations along a stationary surface will have the same stochastic properties regardless of their horizontal locations. Considering this property more closely, one would find it improbable that surfaces can be described by stationary processes [e.g., the stochastic properties of the surface elevation are changing along consistently uphill terrain; [see Figure 5.2(a)]. Therefore, the stationary property of the surface can be achieved if and only if the surface trend has been removed [see Figure 5.2(b)].

An ergodic random process can be formally defined as the process whose time/spatial averaging is equivalent to ensemble averaging. Therefore, the stochastic properties of an ergodic surface at any location can be obtained through the stochastic properties of given elevations at point ensemble, the reference points, along the surface.

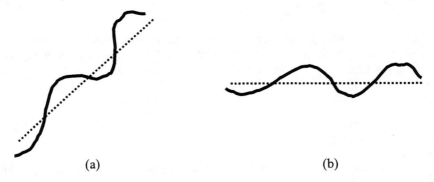

(a)   (b)

Figure 5.2 Part (a) shows a nonstationary surface profile (e.g., the expected elevation increases as we move from left to right); part (b) shows a stationary surface profile where the expected elevation does not change as we move along the surface. Part (b) was obtained by removing the trend (dotted line) from the surface in part (a).

Based on the above definitions, we can obtain the variance-covariance function describing the surface using the given elevations at the reference points if we assume that a stationary and ergodic random process can represent such a surface. Once again stationary and ergodic properties of the surface elevations can be achieved by removing the trend "low-frequency" component of the surface. In other words, we are only dealing with the local fluctuation "high-frequency" component of the surface. The trend can be estimated through the TSA analysis as previously discussed in Chapter 4 (Section 4.4.1.1).

Now the problem at hand is deriving the variance-covariance function from a given set of sampled elevations at the reference points. First, we can determine the variance (covariance between two points that are zero distance apart) as follows

$$C\{0\} = E\left\{ \left(z - E\{z\}\right)_{1 \times n}^{T} \left(z - E\{z\}\right)_{n \times 1} \right\}$$

(5.20)

Here, we are assuming the availability of n reference points. After removing the trend (i.e., $E\{z\} = 0$), the variance $C\{0\}$ can be computed as:

$$C\{0\} = E\left\{ (z)_{1 \times n}^{T} (z)_{n \times 1} \right\}$$

$$C\{0\} = \frac{1}{n} \sum_{i=1}^{n} z_i^2$$

(5.21)

Now the covariance between points that are d-distance apart from each other can be mathematically defined as follows:

$$C\{d\} = E\left\{ \left(z_i - E\{z_i\}\right)_{1 \times n_d}^{T} \left(z_j - E\{z_j\}\right)_{n_d \times 1} \right\}$$

(5.22)

where i and j are the indices of the reference points that are d-distance apart from each other and n_d is the number of these point pairs. Once again, by assuming that the trend has been removed (i.e., $E\{z_i\} = E\{z_j\} = 0$), the covariance $C\{d\}$ can be mathematically given as:

$$C\{d\} = \frac{1}{n_d} \sum_{n_d} z_i z_j$$

(5.23)

In summary, to compute the variance-covariance function $C\{d\}$, we can proceed as follows:

1. Remove the surface trend using trend surface analysis (to guarantee the stationary and ergodic properties of the surface).
2. Compute the variance (covariance at zero distance $C\{0\}$) according to (5.21) using the residual elevations after removing the trend.

3. Select a certain distance d and determine all the point pairs that are d-distance apart from each other (e.g., n_d is the number of point pairs).
4. Compute $C\{d\}$ according to (5.23) using the residual elevations after removing the trend.
5. Repeat steps 3 and 4 after selecting other distances.

The determined variance-covariance function is commonly known as the *empirical variance-covariance function*. The next step involves the selection of a *model variance-covariance function*, which is an analytical function that resembles the empirically determined one [e.g., the Gaussian model covariance in (5.24)]. Using least squares adjustment, we can determine the parameters describing the model variance-covariance function [e.g., C_0 and k in (5.24)].

$$C\{d\} = C_0\, e^{-d^2/(2k^2)} \tag{5.24}$$

After defining the model variance-covariance function, the variance-covariance matrix between the elevations at the reference points can be derived as follows:

$$D\{z_r\}_{n\times n} = \begin{bmatrix} C\{0\} & C\{d_{12}\} & C\{d_{13}\} & \cdots & C\{d_{1n}\} \\ C\{d_{21}\} & C\{0\} & C\{d_{23}\} & \cdots & C\{d_{2n}\} \\ C\{d_{31}\} & C\{d_{32}\} & C\{0\} & \cdots & C\{d_{3n}\} \\ \vdots & \vdots & \vdots & \vdots & \vdots \\ C\{d_{n1}\} & C\{d_{n2}\} & C\{d_{n3}\} & \cdots & C\{0\} \end{bmatrix} \tag{5.25}$$

where d_{ij} is the computed horizontal distance between the ith and jth reference points. On the other hand, the covariance vector between the elevations at the reference and interpolation points can be obtained as follows:

$$C\{z_r, z_p\}_{n\times 1} = \begin{bmatrix} C\{d_{1p}\} \\ C\{d_{2p}\} \\ C\{d_{3p}\} \\ \vdots \\ C\{d_{np}\} \end{bmatrix} \tag{5.26}$$

where d_{ip} is the horizontal distance between the ith reference point and the interpolation point. After computing the dispersion and covariance matrices, we should proceed according to (5.15) to (5.18) to derive an estimate of the height at the interpolation point as well as the associated quality measures.

5.6 KRIGING AS THE BEST LINEAR UNBIASED ESTIMATES (BLUE)

So far, the estimated height at the interpolation points has been derived as a linear combination, weighted sum, of the elevations at the reference points. The weights have been chosen to ensure that the corresponding estimation error has the minimum variance within the class of linear estimates. In addition to the linear and best properties of the estimated height at the interpolation point, one might require that this estimate should be an *unbiased estimate* of the true height at the point in question. Mathematically, this can be defined as follows:

$$E\{e_p\} = E\{z_p\} - w^T E\{z_r\} = 0.0 \tag{5.27}$$

Now, we would like to investigate the necessary condition to assure the unbiased property of the derived estimate. If we are dealing with a stationary surface, then the stochastic properties of the surface (including its expectation) should be same everywhere. Therefore, $E\{z_r\}$ can be defined as follows:

$$E\{z_r\} = K_{n \times 1} E\{z_p\} \tag{5.28}$$

where K is a vector whose elements are all unity (i.e., $K^T_{1 \times n} = \begin{bmatrix} 1 & 1 & 1 & \cdots & 1 \end{bmatrix}$). Accordingly, (5.27) can be rewritten as:

$$E\{e_p\} = \left(1 - w^T_{1 \times n} K_{n \times 1}\right) E\{z_p\}$$

$$E\{e_p\} = \left(1 - \sum_{i=1}^{n} w_i\right) E\{z_p\} \tag{5.29}$$

Therefore, the estimated height at the interpolation point would be unbiased if and only if the summation of the weights is exactly one.

Following the previous discussion, we can achieve a BLUE of the elevation at the interpolation point by modifying the target function in (5.14a) as follows:

$$\varphi(w) = D\{z_p\} - 2 w^T C\{z_r, z_p\} + w^T D\{z_r\} w = \min(w) \,\&\, (w^T K = 1) \tag{5.30}$$

This target function can be modified by including the LaGrange multiplier (λ) as follows:

$$\varphi'(w, \lambda) = D\{z_p\} - 2 w^T C\{z_r, z_p\} + w^T D\{z_r\} w + 2\lambda(w^T K - 1) = \min(w, \lambda) \tag{5.31}$$

The solution to this target function can be obtained in the same way as the solution of the target function in (5.14a):

$$\partial\varphi'/\partial w = -2\,C\{z_r,z_p\} + 2\,D\{z_r\}w + 2\lambda\,K = 0$$
$$\partial\varphi'/\partial\lambda = 2(w^T\,K - 1) = 0 \tag{5.32}$$

This yields the following normal equations:

$$\begin{bmatrix} D\{z_r\}_{n\times n} & K_{n\times 1} \\ K^T_{1\times n} & 0 \end{bmatrix}_{(n+1)\times(n+1)} \begin{bmatrix} w_{n\times 1} \\ \lambda \end{bmatrix}_{(n+1)} = \begin{bmatrix} C\{z_r,z_p\}_{n\times 1} \\ 1 \end{bmatrix}_{(n+1)} \tag{5.33}$$

Finally, the variance of the estimation error can be derived as follows:

$$D\{e_p\} = \left(w^T\,D\{z_r\}\,w\right)_{1\times 1} - \left(2\,w^T\,C\{z_r,z_p\}\right)_{1\times 1} + D\{z_p\}_{1\times 1} \tag{5.34}$$

From (5.33), it can be seen that:

$$D\{z_r\}\,w = C\{z_r,z_p\} - K\lambda \tag{5.35}$$

Substituting (5.35) into (5.34), we get:

$$D\{e_p\} = w^T\,C\{z_r,z_p\} - w^T K\lambda - 2\,w^T\,C\{z_r,z_p\} + D\{z_p\} \tag{5.36}$$

Remember that for an unbiased estimate, $w^T K = 1$. Therefore, (5.36) can be rewritten as follows:

$$D\{e_p\} = D\{z_p\} - w^T\,C\{z_r,z_p\} - \lambda = C\{0\} - w^T\,C\{z_r,z_p\} - \lambda \tag{5.37}$$

Finally, the variance of the interpolated height can be derived according to (5.18) as follows:

$$D\{\hat{z}_p\} = D\{z_p\} - D\{e_p\}$$
$$D\{\hat{z}_p\} = D\{z_p\} - \left[D\{z_p\} - w^T\,C\{z_r,z_p\} - \lambda\right] \tag{5.38}$$
$$D\{\hat{z}_p\} = w^T\,C\{z_r,z_p\} + \lambda$$

The BLUE estimate of the height at the interpolation point is known as *ordinary KRIGING*.

In summary, the BLE/BLUE interpolation methods should proceed as follows:

1. Remove the trend to ensure the stationary and ergodic properties of the surface.
2. Using the measured elevations at the reference points after removing the trend (residual elevations), determine the experimental variance-covariance function $C\{d\}$.
3. Choose a model covariance function that resembles the experimentally determined variance-covariance function.
4. Using least squares adjustment, determine the parameters involved in the model variance-covariance function.
5. Use the provided formulas to determine the interpolated residual height as well as its quality [according to either BLE, (5.15)–(5.18), or BLUE formulations, (5.33) – (5.38)].
6. Add the interpolated residual height to the trend to get the final interpolated elevation.

Similar to the inverse distance weighting (in Section 4.4.2.2), KRIGING interpolation can be performed while considering surface constraints such as break lines. In such a case, the covariance function between the reference and/or interpolation points on opposite sides of a break line should be set to zero. A covariance value of zero indicates no dependence between the elevations, which should be the case when considering points on opposite sides of a break line.

5.7 KRIGING USING SEMIVARIANCE ANALYSIS

Until now, an estimate of the height at the interpolation point has been attained through the use of the variance-covariance function, which represents the interaction between the elevations at nearby points along the surface. Instead of the variance-covariance function, the semivariance could be used to define the spatial dependence between samples (elevations) at various locations. The semivariance $\gamma\{d\}$ can be mathematically described as half the variance of the differences between all possible point pairs that are d-distance apart (Clark, 1979).

$$\gamma\{d\} = \frac{1}{2} E\left\{\left[\left(z_i - z_j\right) - E\left\{z_i - z_j\right\}\right]^2\right\} \tag{5.39}$$

where i and j are the indices of the reference points that are d-distance apart from each other. Assuming that the surface under consideration is a stationary random process (i.e., $E\{z_i\} = E\{z_j\}$), (5.39) can be rewritten as follows:

$$\gamma\{d\} = \frac{1}{2} E\left\{\left[(z_i - z_j)\right]^2\right\}$$

$$\gamma\{d\} = \frac{1}{2n_d} \sum_{n_d} (z_i - z_j)^2 \tag{5.40}$$

where n_d is the number of point pairs that are d-distance apart from each other. The magnitude of the semivariance between points depends on the distance between the points. A smaller distance yields a smaller semivariance while a larger distance results in a larger semivariance. The semivariance at a zero distance is zero (i.e., $\gamma\{0\}$ = zero) since we are comparing the point with itself. The plot of the semivariance versus the distance is known as semivariogram or simply variogram. The expected shape of the variogram can be seen in Figure 5.3. There are two important features that can be observed in the variogram plot; the range/span and the sill. The range can be defined as the distance at which the semivariance will reach a plateau. In other words, the range is the distance beyond which elevations at the corresponding point pairs are not related. The range is important for the interpolation process since it defines the extent of the neighborhood that can be used for local interpolation. The semivariance at a range is defined as the "sill" (i.e., sill $= \gamma\{range\}$).

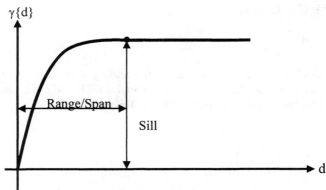

Figure 5.3 Expected shape of the variogram describing the semivariance between the elevations at points that are d-distance apart from each other.

In order to derive the corresponding interpolation formulas when using the semivariance as a measure of the interaction between the elevations at nearby points, we have to establish the mathematical relationship between the semivariance and the variance-covariance function. For stationary and ergodic surfaces, we have established the following mathematical definitions:

$$C\{d\} = \frac{1}{n_d} \sum_{n_d} z_i z_j$$

$$\gamma\{d\} = \frac{1}{2n_d} \sum_{n_d} (z_i - z_j)^2$$

(5.41)

By expanding the expression for the semivariance, we get:

$$\gamma\{d\} = \frac{1}{2n_d} \sum_{n_d} (z_i^2 - 2z_i z_j + z_j^2)$$

$$\gamma\{d\} = \frac{1}{2n_d} \left(\sum_{n_d} z_i^2 + \sum_{n_d} z_j^2 \right) - \frac{1}{n_d} \sum_{n_d} z_i z_j$$

(5.42)

For stationary surfaces:

$$\frac{1}{2n_d} \sum_{n_d} z_i^2 = \frac{1}{2n_d} \sum_{n_d} z_j^2 = \frac{1}{2} C\{0\}$$

(5.43)

which would lead to the final expression as follows:

$$\gamma\{d\} = C\{0\} - C\{d\}$$

(5.44)

Using (5.44), we can derive the corresponding formulas for the simple KRIGING (BLE). Remember that for the simple KRIGING, the interpolated height and the weight vectors are determined as follows:

$$\hat{z}_p = w^T z_r$$

$$w_{n\times1} = [D\{z_r\}]^{-1}_{n\times n} \left[C\{z_r, z_p\} \right]_{n\times1}$$

(5.45)

Substituting (5.44) into (5.45), we can derive the weighting vector.

$$w_{n\times1} = \left[C\{0\} KK^T - \gamma\{z_r, z_r\} \right]^{-1}_{n\times n} \left[C\{0\} K - \gamma\{z_r, z_p\} \right]_{n\times1}$$

(5.46)

where K is an $n\times1$ vector whose elements are all ones. Thus, KK^T is an $n\times n$ matrix whose elements are all ones. Equation (5.46) can be reorganized as follows:

$$\left[C\{0\}\, KK^T - \gamma\{z_r, z_r\}\right]_{n\times n} w_{n\times 1} = \left[C\{0\}\, K - \gamma\{z_r, z_p\}\right]_{n\times 1}$$
$$C\{0\}\, KK^T\, w - \gamma\{z_r, z_r\}\, w = C\{0\}\, K - \gamma\{z_r, z_p\} \tag{5.47}$$

Assume that the interpolated elevation is a nearly unbiased estimate of the true elevation at the interpolation point, then $K^T w \approx 1$. Therefore, (5.47) can be simplified to:

$$C\{0\}\, K - \gamma\{z_r, z_r\}\, w = C\{0\}\, K - \gamma\{z_r, z_p\}$$
$$w_{n\times 1} = \left[\gamma\{z_r, z_r\}\right]_{n\times n}^{-1} \left[\gamma\{z_r, z_p\}\right]_{n\times 1} \tag{5.48}$$

In a similar way, the variance-covariance matrix of the estimation error can be derived as follows:

$$D\{e_p\} = C\{0\} - w^T \left[C\{z_r, z_p\}\right]$$
$$D\{e_p\} = C\{0\} - w^T \left[KC\{0\} - \gamma\{z_r, z_p\}\right]$$
$$D\{e_p\} = C\{0\} - w^T KC\{0\} + w^T \gamma\{z_r, z_p\} \tag{5.49}$$
$$D\{e_p\} = C\{0\} - C\{0\} + w^T \gamma\{z_r, z_p\}$$
$$D\{e_p\} = w^T \gamma\{z_r, z_p\}$$

Finally, the variance of the interpolated height can be derived according to (5.50):

$$D\{\hat{z}_p\} = w^T C\{z_r, z_p\}$$
$$D\{\hat{z}_p\} = w^T \left[K C\{0\} - \gamma\{z_r, z_p\}\right]$$
$$D\{\hat{z}_p\} = w^T K C\{0\} - w^T \gamma\{z_r, z_p\} \tag{5.50}$$
$$D\{\hat{z}_p\} = C\{0\} - w^T \gamma\{z_r, z_p\}$$

In a similar fashion, the formulations for the ordinary KRIGING (BLUE) using the semivariance analysis can be derived:

$$\hat{z}_p = w^T z_r \tag{5.51}$$

where, the weight vector can be derived using (5.52):

$$\begin{bmatrix} \gamma\{z_r, z_p\} & K \\ K^T & 0 \end{bmatrix} \begin{bmatrix} w \\ \lambda \end{bmatrix} = \begin{bmatrix} \gamma\{z_r, z_p\} \\ 1 \end{bmatrix} \tag{5.52}$$

The variance of the estimation error would be given by:

$$D\{e_p\} = w^T \gamma\{z_r, z_p\} + \lambda \tag{5.53}$$

Finally, the variance of the interpolated height can be determined as follows:

$$D\{\hat{z}_p\} = C\{0\} - w^T \gamma\{z_r, z_p\} - \lambda \tag{5.54}$$

In summary, the interpolation using the semivariance analysis would proceed as follows:

1. Ensure the stationary and ergodic properties of the surface by removing the trend.
2. Using the reduced elevations, after removing the trend, at the reference points, determine the experimental variogram $\gamma\{d\}$.
3. Choose a model variogram that resembles the experimentally determined variogram.
4. Using least squares adjustment, determine the parameters involved in the model variogram.
5. Using the above formulas, determine the interpolated residual height as well as its quality.
6. Add the interpolated residual height to the trend to get the final interpolated elevation.

Similar to covariance computation, semivariogram computation can consider break lines within the interpolation procedure. In such a case, the semivariance ($\gamma\{d\}$) between the reference and/or interpolation points on opposite sides of a break line should be set to $C\{0\}$. Such a semivariance corresponds to a covariance value of (0) [refer to (5.44)]. As mentioned earlier, a covariance value of zero implies that there is no dependence between the elevations at the point pair in question.

5.8 NUMERICAL EXAMPLES

5.8.1 Numerical Example 1

Using the given set of reference points in Figure 5.4, determine the empirical variogram and variance-covariance functions. The numbers within parentheses in Figure 5.4 correspond to the point ID and its elevation, respectively.

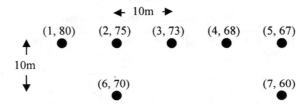

Figure 5.4 Distribution of the points and corresponding elevations.

To ensure the stationary property, the trend has to be removed from the given data. For simplicity, let's assume that the trend can be represented by a zero-order (constant shift) polynomial. The residual elevations after removing the trend (represented by the average height of 70.43m) can be seen in Figure 5.5.

The variance at a zero-distance $(C\{0\})$ can be computed according to (5.21) to be 35.10 m². For the covariance $(C\{d\})$, we have to consider point pairs that are d-distance apart from each other. In this example, we are dealing with seven reference points. Selecting two points at a time produces 21 point-pairs (distances). The 21 distance-pairs can be classified into eight distinct groups. The group number, the corresponding distance, and the necessary computations to evaluate the semivariance and the covariance values are listed in Table 5.1. The empirical semivariance and variance-covariance values are listed in Tables 5.2 and 5.3, respectively. A graphical plot of the variogram and the variance-covariance functions are shown in Figures 5.6 and 5.7, respectively.

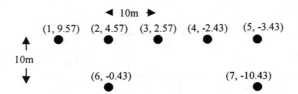

Figure 5.5 Distribution of the points and corresponding elevations after removing trend.

Note that due to the given layout of the reference points in this example, the distances could be grouped into eight distinct groups. It should be expected that we might not be able to have this grouping possibility if we are dealing with randomly distributed reference points. In such a case, the distance range is divided into intervals. Point-pairs whose distances from each other lie within a given interval should be used to compute the corresponding semivariance or the covariance values, which would be assigned to the middle point of this interval.

Table 5.1

Variogram and Covariance Computation

Grp. #	Dist. (m)	# of pairs	C{d}	γ {d}	Pairs	DZ²	Z_iZ_j
1	10	6	15.23	10.75	1,2	25	43.73
					2,3	4	11.74
					3,4	25	-6.25
					4,5	1	8.33
					2,6	25	-1.97
					5,7	49	35.77
2	20	3	1.55	22.33	1,3	49	24.59
					2,4	49	-11.11
					3,5	36	-8.82
3	30	3	-11.49	51.33	1,4	144	-23.26
					2,5	64	-15.68
					6,7	100	4.48
4	40	1	-32.83	84.5	1,5	169	-32.83
5	14.142	3	6.70	28.83	1,6	100	-4.12
					3,6	9	-1.11
					4,7	64	25.34
6	22.361	2	-12.89	43.25	4,6	4	1.04
					3,7	169	-26.81
7	31.623	2	-23.10	58.5	2,7	225	-47.67
					5,6	9	1.47
8	41.231	1	-99.82	200	1,7	400	-99.82

Table 5.2

Empirical/Experimental Variogram

$d\ (m)$	$\gamma\ \{d\}\ (m^2)$
0	0
10	10.75
14.142	28.83
20	22.33
22.361	43.25
30	51.33
31.623	58.5
40	84.5
40.231	200

Table 5.3

Empirical/Experimental Variance-Covariance Function

$d\ (m)$	$C\{d\}\ (m^2)$
0	35.1
10	15.23
14.142	6.70
20	1.55
22.361	-12.89
30	-11.49
31.623	-23.1
40	-32.83
40.231	-99.82

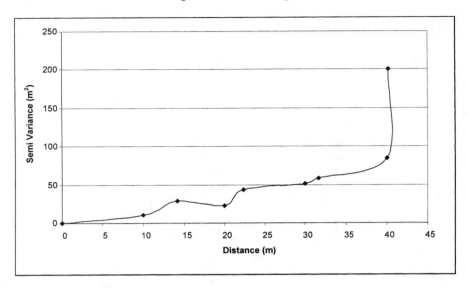

Figure 5.6 Empirical variograms.

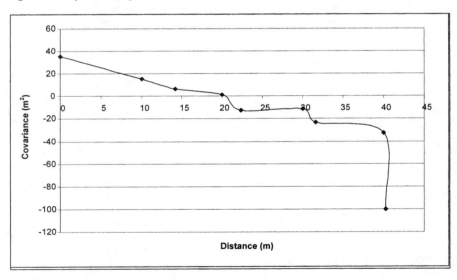

Figure 5.7 Empirical variance-covariance diagram.

5.8.2 Numerical Example 2

In this example, we would like to estimate the elevation at the interpolation point *p* using three reference points. The coordinates of the reference and interpolation points as well as the distances between them are given in Table 5.4. A prior analysis has produced a variogram, which is linear with a slope of 4.0 m^2/km (i.e., $\gamma\{d\} = 4.0d$ m^2, where d is in units of kilometers). Values of the semivariance corresponding to the distances between the points are also given in Table 5.4.

Table 5.4

Coordinates of the Interpolation and Reference Points and Semivariance Computation

Point	East (km)	North (km)	Elevation (m)
1	3.0	4.0	120.0
2	6.3	3.4	103.0
3	2.0	1.3	142.0
p	3.0	3.0	?

Distance between points (km)				
Point	1	2	3	p
1	0	3.3541	2.8792	1.0
2		0	4.7854	3.3242
3				1.9723

Semivariances (m^2)				
	1	2	3	p
1	0	13.4164	11.5168	4.0
2			19.1416	13.2968
3				7.8892

For simple KRIGING, the weighting vector can be determined by solving the following equations:

$$\begin{bmatrix} 0 & 13.4164 & 11.5168 \\ 13.4164 & 0 & 19.1416 \\ 11.5168 & 19.1416 & 0 \end{bmatrix} \bullet \begin{bmatrix} w_1 \\ w_2 \\ w_3 \end{bmatrix} = \begin{bmatrix} 4.0 \\ 13.2968 \\ 7.8892 \end{bmatrix}$$

This yields the following weighting vector:

$$\begin{bmatrix} w_1 \\ w_2 \\ w_3 \end{bmatrix} = \begin{bmatrix} 0.5903 \\ 0.0570 \\ 0.2809 \end{bmatrix}$$

We can observe that the summation of the elements of the weighting vector is close to one $(w_1 + w_2 + w_3 = 0.9282)$. This should be expected since we did not request an unbiased estimate of the elevation at the interpolation points. Another observation is that the first reference point has the highest weight while the second reference point has the smallest weight. This should be expected since the first reference point is the closest point to the interpolation point in question while the second reference point is the farthest. These weights do not depend only on the distances between the reference and the interpolation points but also on the stochastic properties of the surface as represented by the model semivariance function.

Assuming that the surface area in question is stationary (i.e., there is no need to remove the trend), the BLE of the interpolation height and associated variance can be derived as follows:

$$\hat{z}_p = 0.5903 \times 120.0 + 0.0570 \times 103.0 + 0.2809 \times 142.0 = 116.5948 \, \text{m}$$

$$D\{e_p\} = 0.5903 \times 4.0 + 0.0570 \times 13.2968 + 0.2809 \times 7.8892 = 5.3353 \, \text{m}^2$$

For ordinary KRIGING, on the other hand, the weighting vector can be determined by solving the following equations:

$$\begin{bmatrix} 0 & 13.4164 & 11.5168 & 1 \\ 13.4164 & 0 & 19.1416 & 1 \\ 11.5168 & 19.1416 & 0 & 1 \\ 1 & 1 & 1 & 0 \end{bmatrix} \bullet \begin{bmatrix} w_1 \\ w_2 \\ w_3 \\ \lambda \end{bmatrix} = \begin{bmatrix} 4.0 \\ 13.2968 \\ 7.8892 \\ 1.0 \end{bmatrix}$$

This yields the following weighting vector:

$$\begin{bmatrix} w_1 \\ w_2 \\ w_3 \\ \lambda \end{bmatrix} = \begin{bmatrix} 0.6039 \\ 0.0868 \\ 0.3093 \\ -0.7266 \end{bmatrix}$$

We can observe that the summation of the elements of the weight vector is exactly one $(w_1 + w_2 + w_3 = 1.0)$, which is the necessary condition for achieving an unbiased estimate of the elevation at the interpolation point.

Finally, the BLUE of the interpolation and associated variance can be derived as follows:

$$\hat{z}_p = 0.6039 \times 120.0 + 0.0868 \times 103.0 + 0.3093 \times 142.0 = 125.3304\text{m}$$

$$D\{e_p\} = 0.6039 \times 4.0 + 0.0868 \times 13.2968 + 0.3093 \times 7.8892 - 0.7266 = 5.2833\,\text{m}^2$$

Now, let's assume that the interpolation point "p" coincides with the second reference point. For such a scenario, we would like to estimate the height using simple and ordinary KRIGING formulations.

For simple KRIGING, the weighting vector can be determined by solving the following equation:

$$\begin{bmatrix} 0 & 13.4164 & 11.5168 \\ 13.4164 & 0 & 19.1416 \\ 11.5168 & 19.1416 & 0 \end{bmatrix} \bullet \begin{bmatrix} w_1 \\ w_2 \\ w_3 \end{bmatrix} = \begin{bmatrix} 13.4164 \\ 0 \\ 19.1416 \end{bmatrix}$$

Note that the vector on the right-hand side of the above equation has been modified to reflect the semivariances based on the distances between the reference points and the new interpolation point. This yields the following weight vector:

$$\begin{bmatrix} w_1 \\ w_2 \\ w_3 \end{bmatrix} = \begin{bmatrix} 0.000 \\ 1.000 \\ 0.000 \end{bmatrix}$$

The BLE of the interpolation and associated variance can be derived as follows:

$$\hat{z}_p = 0.0 \times 120.0 + 1.0 \times 103.0 + 0.0 \times 142.0 = 103.0\,\text{m}$$

$$D\{e_p\} = 0.0 \times 13.4164 + 1.0 \times 0.0 + 0.0 \times 19.1416 = 0.0\,\text{m}^2$$

For ordinary KRIGING, the weighting vector can be determined by solving the following equation:

$$\begin{bmatrix} 0 & 13.4164 & 11.5168 & 1 \\ 13.4164 & 0 & 19.1416 & 1 \\ 11.5168 & 19.1416 & 0 & 1 \\ 1 & 1 & 1 & 0 \end{bmatrix} \bullet \begin{bmatrix} w_1 \\ w_2 \\ w_3 \\ \lambda \end{bmatrix} = \begin{bmatrix} 13.4164 \\ 0 \\ 19.1416 \\ 1 \end{bmatrix}$$

This yields the following weighting vector:

$$\begin{bmatrix} w_1 \\ w_2 \\ w_3 \\ \lambda \end{bmatrix} = \begin{bmatrix} 0.0000 \\ 1.0000 \\ 0.0000 \\ 0.0000 \end{bmatrix}$$

The BLUE of the interpolation and associated variance can be derived as follows:

$$\hat{z}_p = 0.0 \times 120.0 + 1.0 \times 103.0 + 0.0 \times 142.0 = 103.0\,\text{m}$$

$$D\{e_p\} = 0.0 \times 13.4164 + 1.0 \times 0.0 + 0.0 \times 19.1416 + 0.0000 = 0.0\,\text{m}^2$$

Therefore, we can conclude that KRIGING is an exact interpolation method since the BLE and the BLUE estimates of the interpolated height at the second reference point is exactly the same as the given height of that point. Also, since the interpolation point coincides with one of the reference points, the estimation error has a zero variance (i.e., there is no estimation error).

5.9 DIRECTIONAL VARIOGRAMS/VARIANCE-COVARIANCE FUNCTIONS

So far, we introduced the basics of variance-covariance functions and variograms as geostatistical methods that describe the spatial variations among surface elevations at given distance lags. The variance-covariance function and variogram introduced in the previous sections are *isotropic* (i.e., the behavior of these functions is the same in all directions). Here we are going to explore *directional variance-covariance function/variograms* and the phenomenon of *anisotropy*.

Directional variance-covariance function/variograms define the spatial variation among points separated by a lag vector $\vec{d}$. The difference from the omnidirectional variance-covariance functions/variograms is that $\vec{d}$ is a vector rather than a scalar. For example, if $\vec{d} = \{d_1, d_2\}$, then each pair of compared samples should be separated by d_1 meters in the E-W direction and by d_2 meters in the S-N direction.

In practice, it is difficult to find enough sample points that are separated by exactly the same lag vector $\vec{d}$. Thus, the set of all possible lag vectors is usually partitioned into classes (see Figure 5.8). Vectors that end in the same cell are grouped into one class and the variance-covariance function/variogram values are estimated separately for each class. The number of directions may be different (4, 8, 16, and so on). Variance-covariance function/variograms are estimated using the same equation as the isotropic ones. The only difference here is the points used in the equation are defined as the points located at the tail and head of vector $\vec{d}$.

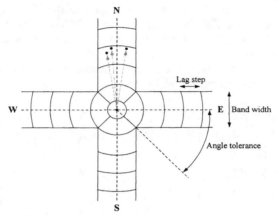

Figure 5.8 Surface partitioning for directional variance-covariance function and variogram computations.

5.10 UNIVERSAL KRIGING

The derived formulas for the simple KRIGING (BLE) and the ordinary KRIGING (BLUE) assume that the surface can be represented by a stationary and ergodic random process. Therefore, these properties have to be ensured by removing the trend from the given elevations at the reference points. The formulas are then applied using the residual elevations. Following the interpolation, the interpolated elevation residual is added to the trend to get the final estimate of the elevation at the interpolation point. The universal KRIGING procedure can be directly applied to the raw data without removing the trend (Davis, 1986). This is due to the fact that the universal KRIGING simultaneously estimates the trend and the weighting vector. Thus, it eliminates the need for the preprocessing step to remove the surface trend.

We will illustrate the formulation of universal KRIGING through an example. Let's assume that we are given 3-D coordinates of five reference points. Using these reference points, we want to estimate the elevation at the location, defined by the horizontal coordinates (X_p, Y_p), of the interpolation point. Let's also assume the availability of the model variogram or model variance-covariance function that describes the interaction between the elevations at various points along the surface. Since the universal KRIGING procedure simultaneously estimates the weighting vector and the trend, a decision has to be made regarding the polynomial representing the trend of the surface. For this example let's assume that the trend can be modeled by a first-order polynomial ($Z = a_0 + a_1 X + a_2 Y$). The constant term of the trend (a_0) represents a constant shift in the Z-direction. Therefore, one can argue that removing or keeping such a term would not affect the stationary property of the surface. In other words, the stationary property can be achieved by just removing the tilt components in the X and Y directions ($a_1 X$ and $a_2 Y$, respectively). Therefore, the universal KRIGING will estimate the five weights associated with the five reference points as well as the tilt components of the trend (a_1 and a_2).

The unknown parameters can be determined by solving the following normal matrix derived using a model variogram:

$$
\begin{bmatrix}
\gamma\{0\} & \gamma\{d_{12}\} & \gamma\{d_{13}\} & \gamma\{d_{14}\} & \gamma\{d_{15}\} & 1 & X_1 & Y_1 \\
\gamma\{d_{21}\} & \gamma\{0\} & \gamma\{d_{23}\} & \gamma\{d_{24}\} & \gamma\{d_{25}\} & 1 & X_2 & Y_2 \\
\gamma\{d_{31}\} & \gamma\{d_{32}\} & \gamma\{0\} & \gamma\{d_{34}\} & \gamma\{d_{35}\} & 1 & X_3 & Y_3 \\
\gamma\{d_{41}\} & \gamma\{d_{42}\} & \gamma\{d_{43}\} & \gamma\{0\} & \gamma\{d_{45}\} & 1 & X_4 & Y_4 \\
\gamma\{d_{51}\} & \gamma\{d_{52}\} & \gamma\{d_{53}\} & \gamma\{d_{54}\} & \gamma\{0\} & 1 & X_5 & Y_5 \\
1 & 1 & 1 & 1 & 1 & 0 & 0 & 0 \\
X_1 & X_2 & X_3 & X_4 & X_5 & 0 & 0 & 0 \\
Y_1 & Y_2 & Y_3 & Y_4 & Y_5 & 0 & 0 & 0
\end{bmatrix}
\begin{bmatrix}
w_1 \\ w_2 \\ w_3 \\ w_4 \\ w_5 \\ \lambda \\ a_1 \\ a_2
\end{bmatrix}
=
\begin{bmatrix}
\gamma\{d_{1p}\} \\ \gamma\{d_{2p}\} \\ \gamma\{d_{3p}\} \\ \gamma\{d_{4p}\} \\ \gamma\{d_{5p}\} \\ 1 \\ X_P \\ Y_P
\end{bmatrix}
\quad (5.55)
$$

In (5.55), we included the LaGrange multiplier (λ) to ensure an unbiased estimate of the elevation at the interpolation point. Finally, the interpolated elevation can be determined as the weighted sum of the raw elevations at the reference points.

In summary, the weights to be applied to the known reference points as well as the coefficients of the trend are simultaneously determined within the universal KRIGING procedure. Since more terms are being estimated, a large sample of reference points is required. Also, a reasonable model variogram or variance-covariance function is assumed and then compared to the empirical variogram or variance-covariance function derived from the residual elevations after removing the trend. If the two are the same, then the assumptions made about the model variograms or the variance-covariance functions as well as the trend are correct, and the unknown value can be determined. If they differ, then another model variogram/variance-covariance function and/or trend must be used until the model and empirical variograms are the same.

5.11 KRIGING: FINAL REMARKS

Based on the discussion regarding various interpolation methods, one can say that KRIGING has a number of advantages over most other interpolation methods. In theory, this is the only interpolation method discussed so far where you can specify the nature of the interpolated height (e.g., BLUE of the true elevation at the interpolation point). In practice, the validity of the interpolated height depends on the correct specification of the parameters describing the model semivariance or the model variance-covariance function. Also, whether the stationary and ergodic properties of the surface has been achieved or not plays a significant role in the quality of the interpolated height (i.e., the trend has been completely removed). However, due to the robustness of the KRIGING procedure — even with a naive selection of parameters — the method will do no worse than conventional local interpolation procedures. Perhaps most importantly, KRIGING yields estimates of the likely error (in the form of standard errors or error variances) at every interpolation point. These error estimates can be mapped to give a direct assessment of the reliability of the interpolated surface. Since the estimated weights for the reference points are determined using the stochastic properties of the surface (e.g., either through the semivariance or the variance-covariance functions), the neighborhood to be used for interpolation will be adapted according to the nature of the surface in question. For other interpolation methods, the neighborhood for interpolation is empirically determined.

The price that must be paid for the above benefits is the increased computational complexity. A large set of simultaneous equations must be solved for every interpolation point estimated by KRIGING. Therefore, computer run time will be significantly longer if a map is produced by KRIGING rather than by conventional interpolation methods. In addition, an extensive prior study of the data must be made to test whether the surface is stationary or not and to determine the form of the model variogram/variance-covariance function. These factors are not independent. Therefore, trial-and-error experimentation might be necessary to determine the best combination. For this reason, to justify the additional costs of analysis and processing, KRIGING should be applied in those instances where the best possible estimates of the surface are essential, the data is of reasonably good quality, and estimates of the error are needed.

References

Brown, R., and Hwang, P., *Introduction to Random Signals and Applied Kalman Filtering*, New York: John Wiley & Sons, 1992.

Clark, I., *Practical Geostatistics*, London: Applied Science Publishers LTD, 1979.

Davis, J., *Statistics and Data Analysis in Geology*, New York: John Wiley & Sons, 1986.

Gelb, A. (Ed.), *Applied Optimal Estimation*, Cambridge, MA: M.I.T. Press, 1979.

Koch, K., *Parameter Estimation and Hypothesis Testing in Linear Models*, Berlin: Springer-Verlag, 1987.

Mikhail, E., *Observations and Least-Squares*, New York: Thomas Y. Crowell, 1976.

Mikhail, E., and Gracie, G., *Analysis and Adjustment of Survey Measurements*, New York: Van Nostrand Reinhold Company, 1981.

Wolf, P., and Ghilani, C., *Adjustment Computations: Statistics and Least Squares in Surveying and GIS*, New York: John Wiley & Sons, 1997.

Chapter 6

DTM Generalization and Quality Control

6.1 INTRODUCTION

In the previous two chapters, we discussed possible alternatives for retrieving the continuous surface of the Earth from discrete samples as represented by the elevations at the reference points. In this chapter, we will discuss other aspects of DTM manipulation; namely, DTM generalization and quality control. DTM generalization is a filtering process, where the elevation data is reduced to produce a subset of the original data while maintaining key surface characteristics. The quality control investigates the factors that might affect the accuracy of the DTM data as well as the quantitative measures for assessing that accuracy. The quality control analysis starts by investigating error sources in the collected elevation data from different acquisition systems. This discussion will be followed by an outline of several techniques for comparing overlapping elevation data, which might have been captured at different times by the same or different acquisition systems, for quality control purposes.

6.2 DTM GENERALIZATION

DTM data is usually used with other data and/or information to provide new products and information. For example, DTM data is combined with aerial and space imagery to produce orthophotos of the area in question. Also, DTM data can be integrated with vegetation cover, soil type, and meteorological conditions to provide information regarding the potential of fire, landslides, or flooding. To achieve the aforementioned objectives, DTM and other data sources are usually incorporated in a GIS database (Bossler et al., 2002). The individual components of a GIS database are available at different scales and are usually cross-referenced to each other. Real-time processing and visualization of these components require having several scaled versions of the involved data with different levels of detail (i.e., different generalization levels of the involved data including the DTM). In general, the generalization process aims at filtering the original data to obtain a reduced version while maintaining the key characteristics of the original data (e.g., drainage channels, ridges, valleys, pits, peaks, passes, and other surface discontinuities). The nature of the generalization techniques depends on the DTM representation methodology (e.g., regularly spaced data and triangular irregular networks), which will be discussed in the following sections.

6.2.1 Generalization of Regularly Spaced Data

A DTM represented by regularly spaced data in a matrix form is usually denoted as a digital elevation model (DEM). DEM generalization procedures can be grouped into two major categories. The first category of procedures does not change the number of the involved data points. The second category encompasses algorithms that eliminate redundant data points based on some criterion. The former category is known as global filters, which can be applied either in the spatial or frequency domain. The second category of DEM generalization techniques is known as selective filters.

Global filtering techniques are used to smooth or emphasize terrain characteristics. Smoothing filters remove terrain details and are called low-pass filters since they remove the high-frequency components of the terrain that correspond to surface details, while maintaining the low-frequency components that define the global appearance of the terrain. For example, Figure 6.1 illustrates smoothing filters applied to the elevation data in the spatial domain. On the other hand, Figure 6.2 depicts the smoothing operation in the frequency domain. Enhancing filters emphasize the terrain details and are known as high-pass filters. In general, enhancing filters are rarely used for DTM generalization. However, they can be used to emphasize high-frequency components of the terrain such as fault lines and characteristic surface points (Figure 6.3). Global filtering techniques can be applied in the spatial domain using convolution (Figure 6.1), or in the frequency domain using Fourier or wavelet transformation analysis (Figures 6.2 and 6.3).

Selective filtering techniques are mainly used to eliminate points that do not carry significant information about the terrain. Thus, selective filtering procedures will change the regular structure of the original DEM to yield a structure similar to that of a quad tree (Figure 6.4). The significance of the terrain points can be evaluated by comparing original data points and interpolated ones using the neighboring points. If the difference between the original and interpolated points is below a certain threshold value, then the original data point is considered to be redundant and can be eliminated.

(a) (b)

(c) (d)

Figure 6.1 DEM smoothing in the spatial domain: (a) original DTM, and (b), (c), and (d) smoothed versions through spatial convolution with 3×3, 5×5, and 9×9 moving average, respectively.

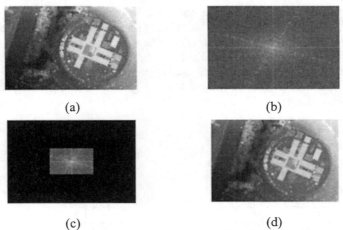

(a) (b)

(c) (d)

Figure 6.2 DEM smoothing in the frequency domain: (a) original DEM, (b) corresponding spectrum, (c) spectrum after suppressing high-frequency components, and (d) corresponding smoothed DEM.

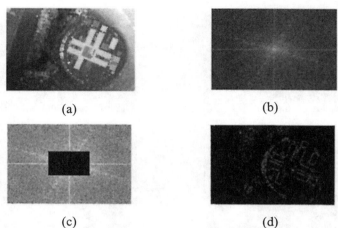

(a)   (b)

(c)   (d)

Figure 6.3 DEM enhancement in the frequency domain: (a) original DEM, (b) corresponding spectrum, (c) spectrum after suppressing low-frequency components, and (d) corresponding enhanced DEM.

Selective terrain generalization procedures can be applied through either subtractive or additive filters. Subtractive filters start with a dense grid, seen in Figure 6.4(a), and proceed by eliminating points with the least significance in an iterative procedure to generate the quad-tree structure in Figure 6.4(b). On the other hand, additive filters start with a coarse grid and add points to the initial grid if these points are crucial for describing the terrain characteristics. In other words, for relatively flat areas, the original coarse grid is maintained. On the other hand, for rugged portions, supplementary surface points will be added. The additive selective filtering procedure resembles the progressive sampling collection mechanism, which is usually implemented by analytical photogrammetric plotters.

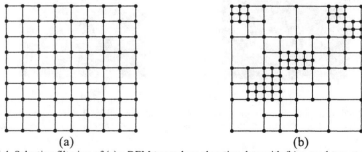

(a)   (b)

Figure 6.4 Selective filtering of (a) a DEM to produce elevation data with (b) a quad-tree structure.

6.2.2 Generalization of Irregularly Spaced Data

The alternative methodology for storing elevation data is triangular irregular networks (TIN), which are constructed by connecting irregularly distributed points covering the area of interest to form a set of nonoverlapping triangles. Compared to regularly spaced data in a DEM, a TIN structure has fewer points. Also, structural features of the terrain can be easily incorporated in a TIN and represented by the triangle legs. Due to the irregular nature of a TIN, global filtering techniques are difficult to apply. On the other hand, the selective filtering techniques, which have been mentioned in the previous section, could be applied.

6.3 DTM QUALITY CONTROL

Having discussed DTM manipulation methodologies, we will shift our focus toward investigating possible errors in collected DTM data. Understanding the error sources in a DTM is essential for improving its quality. Error sources in the collected elevations depend on the DTM acquisition methodology. In this section, we will discuss error sources in derived DTM data from the following sources:

1. Digitizing contours from existing cartographic data;
2. Photogrammetric processing of aerial and satellite imagery;
3. Light detection and ranging (LIDAR) systems.

The following paragraphs will outline the major error sources in elevation data acquired by the aforementioned methodologies. This discussion will be followed by various alternatives for comparing overlapping DTM data to check for any discrepancies resulting from inaccurate acquisition systems and/or changes between the moments of capture associated with temporal datasets.

Digitizing contours from existing cartographic data: Existing cartographic maps were one of the most economic and popular sources for the acquisition of elevation data. The elevations can be derived by manually digitizing contour lines in existing paper maps (Figure 6.5). Alternatively, the contours can be automatically extracted from scanned maps. Errors in the collected elevations would include: (1) existing errors in outdated maps; (2) introduced errors during the scanning/digitization process; and (3) errors in the automatically identified contour lines. Due to technical advances in modern mapping systems (e.g., airborne and spaceborne imaging systems, LIDAR, and RADAR), direct acquisition of accurate elevation information from current spatial data is becoming more affordable. As a result, digitizing existing cartographic data has become a less popular source of collecting elevation data.

Figure 6.5 DTM generation by digitizing existing contour maps.

Photogrammetric processing of aerial and satellite imagery: The acquisition of elevation data from overlapping imagery has become one of the most economic sources for DTM generation, especially with the increasing availability of captured imagery from high resolution imaging satellites (e.g., Ikonos, SPOT, Quickbird, and Eros-A1). The involved steps in the acquisition of elevation data from stereo-images can be summarized as follows (Figure 6.6):

1. Select a specific point in the left image (e.g., point a in Figure 6.6).
2. Identify the corresponding/conjugate point in the right image (e.g., point a´ in Figure 6.6).
3. Use the internal and external characteristics of the imaging system as well as the relevant mathematical model to derive the 3-D coordinates of the corresponding object point (e.g., point A in Figure 6.6).

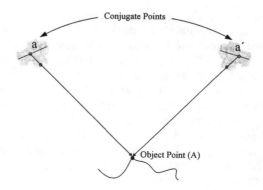

Figure 6.6 DTM generation from stereo-imagery.

For more details regarding photogrammetric principles and DEM generation, refer to Kraus (1993), Schenk (1999), and Mikhail et al. (2001). Error sources in derived object points from photogrammetric techniques include:

1. Errors in the internal characteristics of the implemented imaging systems: The internal characteristics include the principal distance, principal point coordinates, radial lens distortions, and de-centering lens distortions. These characteristics are determined through calibration procedures. Inaccurate determination of these characteristics would lead to errors in the derived elevations.

2. Errors in the external characteristics of the implemented imaging system: The external characteristics include the position and the attitude/orientation of the imaging system relative to the ground coordinate system. The external characteristics are determined through a georeferencing procedure. The georeferencing parameters can be indirectly estimated using ground control points or directly measured using GPS/INS units onboard the imaging platform. The quality of the estimated external characteristics from indirect georeferencing procedures depends on the accuracy and the distribution of the available ground control points. On the other hand, the quality of direct georeferencing measurements depends on the implemented GPS/INS units and the accuracy of the bore-sighting parameters between the GPS/INS units and the implemented camera. Regardless of the georeferencing methodology, errors in the external characteristics will propagate into errors in the derived object points.

3. Errors associated with the image coordinate measurements of conjugate points in overlapping stereo-images: Conjugate points can be manually identified. In such a case, the quality of the image coordinate measurements depends on the geometric resolution of the imaging system as well as the radiometric properties of acquired imagery, which would affect the operator's ability to accurately identify conjugate points. Modern DTM generation techniques from stereo-imagery are based on automatically identified conjugate points in overlapping imagery (Haralick and Shapiro, 1993). The quality of the matching results depends on the nature of the involved scenes. For example, foreshortening and occlusions in high resolution imagery over urban areas will affect the reliability of the matching results. As it can be seen in Figure 6.7, the same vertical facet of the building in stereo-images will appear differently due to the foreshortening problem arising from different perspective views. In addition, repetitive patterns in overlapping imagery would lead to erroneous matching results (Figure 6.8). With regard to the radiometric properties and their effect on feature selecting, it is worth

mentioning that the large dynamic range (radiometric resolution) of digital cameras helps in better identification of features.

4. Errors in the utilized mathematical model relating corresponding image and ground coordinates: Traditionally, rigorous mathematical models, which accurately describe the imaging process of the implemented sensor, have been used to relate corresponding image and ground coordinates. These models require the availability of the internal and external characteristics of the involved sensor. However, some spaceborne image providers do not furnish the sensor characteristics (e.g., Ikonos imagery). In such a case, the users are forced to utilize approximate sensor models [e.g., rational function model (RFM) – or parallel projection model] to relate corresponding image and ground coordinates. The quality of the derived object space from approximate sensor modeling is usually inferior when compared to that derived from the rigorous sensor model. However, approximate sensor models have fewer requirements for ground control. Moreover, they do not call for a thorough understanding of the imaging geometry of the implemented sensor.

Figure 6.7 Foreshortening and occlusions in overlapping imagery.

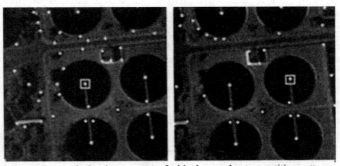

Figure 6.8 Erroneously matched points, centers of white boxes, due to repetitive patterns.

Light detection and ranging (LIDAR) systems: Recent technical advances in direct georeferencing methodologies (e.g., GPS/INS units) are giving rise to the continuous profusion of LIDAR systems for capturing elevation data (Mikhail et al., 2001). LIDAR technology has been conceived to directly and accurately capture digital surfaces. LIDAR systems supply dense geometric surface information in the form of nonselective points. An appealing feature of the LIDAR output is the direct acquisition of 3-D coordinates of object space points. However, there is no inherent redundancy in reconstructed surfaces from LIDAR systems. Therefore, the quality of the derived information depends on the accuracy and validity of the calibration parameters of the different components comprising the LIDAR system. Another characteristic of LIDAR surfaces is that they are mainly positional. In other words, it is difficult to derive semantic information regarding the captured surfaces (e.g., material and type of observed structures). Equation (6.1) describes the mathematical derivation of the ground coordinates of an object point from the different measurements associated with a LIDAR system (Figure 6.9). As per Figure 6.9, a LIDAR system involves four coordinate systems, which are:

1. The ground coordinate system, which is user-defined.
2. The IMU body frame with its origin at the phase center of the GPS antenna.
3. The laser-unit coordinate system with its origin at the laser firing point. The x-axis of the laser-unit coordinate system can be defined to coincide with the flight direction while the z-axis points up.
4. The laser-beam coordinate system with its origin at the laser firing point. The z-axis of the laser-beam coordinate system coincides with the laser beam.

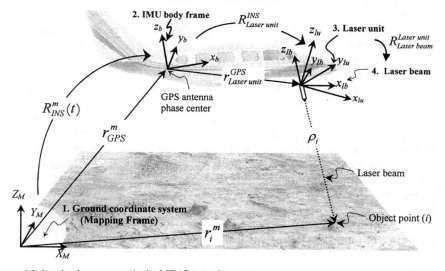

Figure 6.9 Involved parameters in the LIDAR equation.

$$r_i^m = r_{GPS}^m + R_{INS}^m(t)\,R_{laser\ unit}^{INS}\,r_{laser\ unit}^{GPS} + R_{INS}^m(t)\,R_{laser\ unit}^{INS}\,R_{laser\ beam}^{laser\ unit}\begin{bmatrix} 0 \\ 0 \\ -\rho_i \end{bmatrix} \qquad (6.1)$$

where:

r_i^m is the ground coordinates vector of the object point under consideration.

r_{GPS}^m is the ground coordinates vector of the phase center of the GPS antenna.

$r_{laser\ unit}^{GPS}$ is the spatial offset vector between the laser firing point and the phase center of the GPS antenna relative to the laser-unit coordinate system.

$R_{INS}^m(t)$ is the rotation matrix relating the IMU body frame and the ground coordinate system at time t.

$R_{laser\ unit}^{INS}$ is the rotation matrix between the IMU body frame and the laser-unit coordinate system.

$R_{laser\ beam}^{laser\ unit}$ is the rotation matrix between the laser-unit and laser-beam coordinate systems.

ρ_i is the measured range by the LIDAR system to the object point.

According to (6.1), the error sources in the derived coordinates from a LIDAR system can be listed as follows:

1.   Errors in the direct georeferencing parameters provided by the GPS/INS units.
2.   Errors in the spatial and rotational offsets between the various system components. These offsets are usually denoted as the bore-sighting parameters and are determined from a calibration procedure.
3.   Errors in the measured laser range (ρ_i).
4.   Errors in the measured orientation angles relating the laser-beam and laser-unit coordinate systems.

Having discussed possible error sources in captured DTM data from different acquisition systems, we will dedicate the following sections to discuss possible alternatives for assessing the quality of collected elevation data. Throughout the following discussion, the quality of elevation data will be evaluated by comparing two versions of the DTM, one of which will be considered to be the ground truth.

6.3.1 Quality Control in the Spatial Domain

As it mentioned earlier, the quality of DTM data will be evaluated by comparing a given DTM with another version, which can be considered to be the ground truth. Such a comparison can be done using one of the following methodologies:

1. The quality of the DTM can be evaluated by comparing the elevation data at the same geographic location. In this case, the first DTM is interpolated to produce data points coinciding with these in the second DTM. Then, the elevations at corresponding points are compared [e.g., using root mean square error analysis (RMSE)] to provide a quantitative measure of the differences between the two datasets.

2. The second methodology starts by interpolating both DTM versions into a regular grid. Then, corresponding grid posts are compared to derive a quantitative measure of the discrepancies between the two datasets. The main advantage of dealing with regularly spaced DTM data is the possibility of representing the DTM with an image, where the gray values represent elevations (Figure 6.10). Interpolated/gridded DTM data can be compared through the cross-correlation coefficient in (6.2), which expresses the similarity between the respective elevations:

$$\rho = \frac{\sum_{i=1}^{N}\sum_{j=1}^{M}[g_1(x_i,y_j) - \bar{g}_1][g_2(x_i,y_j) - \bar{g}_2]}{\sqrt{\sum_{i=1}^{N}\sum_{j=1}^{M}[g_1(x_i,y_j) - \bar{g}_1]^2 \sum_{i=1}^{N}\sum_{j=1}^{M}[g_2(x_i,y_j) - \bar{g}_2]^2}} \qquad (6.2)$$

where N, M are the number of rows and columns in the elevation models, $g_1(x_i,y_j)$ and $g_2(x_i,y_j)$ are the elevations at the location (x_i,y_j), and $\bar{g}_1$ and $\bar{g}_2$ are the mean elevation values of both elevation models. The closer the correlation coefficient is to positive unity, the closer the two elevation models are to being similar.

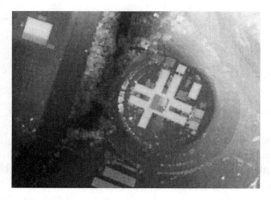

Figure 6.10 Elevation data represented as an image.

6.3.2 Quality Control in the Frequency Domain

Regularly spaced spatial signals such as $g(l_1, l_2)$ can be transformed into the frequency domain using Fourier transformation, (6.3), where $G(k_1, k_2)$ correspond to the amplitudes of various frequency components present in the spatial data (Schalkoff, 1989). To derive the coefficients in (6.3), irregularly distributed elevations have to be interpolated to a regular common grid.

$$G(k_1, k_2) = \sum_{l_2 = 1l_1 = 1}^{N_2} \sum_{}^{N_1} g(l_1, l_2) \, e^{i2\pi(\frac{k_1 l_1}{N_1} + \frac{k_2 l_2}{N_2})} \qquad (6.3)$$

Afterward, transformed elevation data into the frequency domain can be used to evaluate the differences between two DTM datasets by comparing the corresponding amplitudes. The comparison can be carried out by deriving the individual similarity measures in (6.4) for the various frequency components present in both DTM signals.

$$H(k_1, k_2) = G(k_1, k_2)_{DTM_1} \Big/ G(k_1, k_2)_{DTM_2} \qquad (6.4)$$

If the two DTM datasets are equivalent, the ratio in (6.4) should be exactly one. However, differences between the two datasets will make this ratio different from one. Moreover, it is expected that the similarity measures for lower frequency components will be closer to one when compared to those associated with higher frequency components (since low frequency components correspond to the general trend in the elevation data, which is expected to be similar for both datasets). An overall similarity measure between the two datasets can be derived as in (6.5). The smaller the computed value for (σ^2), the more similar the two DTM datasets are. For more details on this topic, interested readers can refer to Terei (1999).

$$\sigma^2 = \frac{1}{2} \sum_{k_1 = 1k_2 = 1}^{N_1} \sum_{}^{N_2} [1 - H(k_1, k_2)]^2 \, G^2(k_1, k_2)_{DTM_1} \qquad (6.5)$$

6.3.3 Quality Control Using Orthophotography

The degree of similarity between two elevation datasets can be evaluated by comparing derived orthophotos from these datasets. Orthophotos, which will be discussed in Chapter 7, are processed images that have the same characteristics of a map; namely, orthogonally projected imagery with a uniform scale (Wolf and Dewitt, 2000). Correctly generated orthophotos will depict the involved objects in their true geographic location within the object space. The quality control procedure requires having two DTM datasets and digital image(s) over the same area. Moreover, the internal (i.e., principal point coordinates, principal distance, and lens distortion parameters associated with the implemented camera) and external (i.e., position and orientation parameters of the imaging platform) characteristics of the involved imagery should be available. The conceptual basis and procedure for utilizing orthophotos for DTM quality control can be summarized as follows:

1. Interpolate the elevation datasets to a common grid.
2. Use the interpolated elevation datasets together with the external and internal characteristics of the imagery to produce two overlapping orthophotos of the area in question (Figure 6.11).
3. Check the discrepancy between the generated orthophotos in the previous step. If a single image has been used, the derived orthophotos will be compatible if and only if the two DTM datasets are identical. Therefore, discrepancies between the generated orthophotos can be attributed to differences between the involved DTM datasets. It is worth mentioning that discrepancies between the derived orthophotos can be evaluated by a simple subtraction of the rectified imagery whenever a single image is used to generate the orthophotos.

If two different images are used to derive the orthophotos, which is the case depicted in Figure 6.11, then radiometric differences between the orthophotos should be expected. Note that geometric differences between the original imagery due to different imaging geometry and sensor view points are compensated for by the orthorectification process. Due to the radiometric differences between the derived orthophotos, the comparison procedure would be more complicated than a simple subtraction. For example, correlation matching (similar to what has been discussed in Section 6.3.1) and/or gray-value normalization (ensuring similar statistical properties — mean and standard deviation — for the involved orthophotos) could be used to compare the resulting orthophotos. However, the matching procedure between the orthophotos will be much easier than matching corresponding features in the original imagery. This is attributed to the fact that the orthophoto rectification eliminates geometric differences between the involved imagery. In other words, there will be no displacements between conjugate features except for problem areas in the available DTM data. The above comparison will reflect the quality of the similarity between the involved DTM data if the following assumptions are true:

1. The interpolation procedure of both DTM datasets into a common grid does not produce any errors. Note that this assumption is rarely met, especially when dealing with large-scale data over urban areas.
2. The internal characteristics of the imaging system are valid.

3. The external characteristics of the involved imagery are valid.
4. There are no changes in the object space between the involved DTM data and/or imagery. In other words, detected discrepancies (mismatches between the orthophotos) are only attributable to problem areas in the DTM data.

In addition to the previous quantitative measures for evaluating discrepancies between the generated orthophotos, we can visually compare them to qualitatively evaluate their similarity. For example, smooth continuity along linear features (e.g., building and road boundaries) indicates a good compatibility between these orthophotos (Figure 6.11).

Figure 6.11 DTM quality control by comparing overlapping orthophotos.

6.3.4 Automatic Surface Matching for Quality Control

The above-mentioned similarity measures require prior interpolation of both DTM datasets to a common grid. The interpolated elevations at corresponding grid nodes are then compared, either directly (Section 6.3.1) or indirectly (Sections 6.3.2 and 6.3.3), for quality control analysis. It should be noted that the interpolation procedure will introduce some errors, especially when dealing with high resolution elevation data in urban areas. The interpolation errors will negatively affect the derived quality control measures. Therefore, alternative measures need to be developed, which can deal with the available elevation data in its raw format. The following paragraphs present an alternative approach for evaluating the degree of similarity between two elevation datasets, which are represented by an irregularly distributed set of points. Moreover, the datasets can be given relative to different reference frames. Finally, this approach does not assume one-to-one correspondence between the involved points. In summary, the presented approach will evaluate the transformation parameters between the involved reference frames, determine the correspondence between conjugate surface elements, and derive an estimate of the degree of similarity between the two datasets.

To illustrate the conceptual basis of the suggested approach, let us assume that we are given two sets of irregularly distributed points that describe the same surface (Figure 6.12). Let $S_1 = \{p_1, p_2, ..., p_l\}$ be the first set and $S_2 = \{q_1, q_2, ..., q_m\}$ be the second set where $l \neq m$. These points are randomly distributed and the correspondences between them are not known. Furthermore, it cannot be assumed that there is one-to-one correspondence between the two point datasets. Also, the two point sets might be given relative to two reference frames. The transformation between these reference frames is assumed to be modeled by a seven-parameter transformation involving three shifts, one scale, and three rotation angles (X_T, Y_T, Z_T, S, ω, φ, and κ, respectively). Hence, the problem at hand is to determine the degree of similarity between the two point sets describing the surface, establish the correspondences between conjugate surface elements, and estimate the transformation parameters between the respective reference frames. The proposed approach creates a TIN model using the points in S_1 to form a group of non-overlapping triangles (Figure 6.12). The individual triangles in the derived TIN are assumed to represent planar patches.

Now, let us consider the surface patch S_p in S_1, which is defined by the three points p_a, p_b, and p_c. The fact that a point q_i in the second set S_2 belongs to the surface patch S_p in the first set can be mathematically described by the constraint in (6.6) (Figure 6.12).

$$\begin{vmatrix} x_{q_i'} & y_{q_i'} & z_{q_i'} & 1 \\ x_{p_a} & y_{p_a} & z_{p_a} & 1 \\ x_{p_b} & y_{p_b} & z_{p_b} & 1 \\ x_{p_c} & y_{p_c} & z_{p_c} & 1 \end{vmatrix} = 0.0 \tag{6.6}$$

where $(x_{q_i'}, y_{q_i'}, z_{q_i'})$ in the above equation represent the transformed point coordinates from the second dataset into the reference frame associated with (6.7). Equation (6.6) simply states that the volume defined by the points q_i', p_a, p_b, and p_c is zero. In other words, these points are coplanar (i.e., the normal distance between q_i', and the surface patch S_p is zero). After establishing the correspondences between the points in S_2 and the patches in S_1, one can solve for the transformation parameters (implicitly considered in the first row of the determinant in (6.6) through the substitution of (6.7)) using a least squares adjustment procedure. In this regard, the estimated variance component (σ_o^2) from the adjustment procedure represents the goodness of fit (degree of similarity) between the two point sets after the coalignment of the respective reference frames. Close similarity between the DTM datasets will be reflected by a small variance component (σ_o^2).

$$
\begin{bmatrix} x_{q_i'} \\ y_{q_i'} \\ y_{q_i'} \end{bmatrix} = \begin{bmatrix} X_T \\ Y_T \\ Z_T \end{bmatrix} + S\,R_{(\omega,\,\varphi,\,\kappa)} \begin{bmatrix} x_{q_i} \\ y_{q_i} \\ z_{q_i} \end{bmatrix} \tag{6.7}
$$

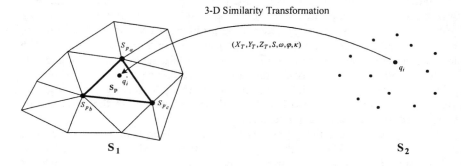

Figure 6.12 Comparing two datasets of irregularly distributed sample points that describe the same surface.

So far, we have established the mathematical model, which can be used to derive the transformation parameters between the reference frames associated with the point sets. Also, we derived a quantitative measure σ_o^2, which describes the degree of similarity between the two point sets. However, the derivation of these estimates requires the knowledge of the correspondence between conjugate surface elements in the available datasets (i.e., which points in S_2 belong to which patches in S_1). Figuring out such a correspondence is difficult since the two sets are given relative to different reference systems. Thus, the remaining problem is to establish a reliable procedure for identifying such a correspondence. The proposed solution to this intricate problem is based on a voting scheme, which simultaneously establishes the correspondence between conjugate surface elements and the transformation parameters between the two datasets.

The suggested voting scheme will determine the transformation parameters without knowing the correspondence between points in the second set and conjugate patches in the first one. As mentioned earlier, the parameters of the 3-D similarity transformation in (6.7) can be estimated once the correspondence between seven points in S_2 and seven patches in S_1 is known, leading to seven constraints of the form in (6.6). The suggested procedure can start by choosing any seven points in S_2 and matching them with all possible surface patches in S_1 leading to several matching hypotheses. For each matching hypothesis, a set of seven equations can be written and used to solve for the transformation parameters. Repeat this procedure until all possible matches between the points in S_2 and the patches in S_1 are considered. It should be obvious that correct matching hypotheses will lead to the same parameter solution. Therefore, the most frequent solution resulting from the previous matching hypotheses will be the correct transformation parameters relating the two datasets in question. Also, the matching hypotheses that led to this solution constitute the correspondences between the points in S_2 and the patches in S_1. A seven-dimensional accumulator array, which is a discrete tessellation of the expected solution space, can be used to keep track of the matching hypotheses and the associated solutions. The correct solution will manifest itself as a peak in that accumulator array. However, exploring all possible matching hypotheses will lead to a combinatorial explosion. For example, if we want to consider all independent combinations between S_1 and S_2, the total number of solutions is $s = \dfrac{(nm)!}{7!\,(nm-7)!}$, where n is the number of patches in S_1 set and m is the number of points in S_2. Moreover, the memory requirement of a 7-D accumulator array is impractical. This problem is caused by our attempt to simultaneously determine the seven transformation parameters.

To avoid the above-mentioned problems, the parameters can be sequentially and individually solved for in an iterative manner starting from some approximate values. Consequently, the accumulator array becomes one-dimensional and the memory requirement problem disappears. The total number of point-to-surface patch combinations reduces to $m \times n$. The suggested procedure can be summarized as follows:

1. Establish some approximate values for the unknown parameters.
2. Select one of the parameters (e.g., X_T). Initialize the corresponding 1-D accumulator array to zero votes. The values of the other parameters (Y_T, Z_T, S, ω, φ, and κ) are considered to be constants.
3. Assume a matching hypothesis between a point q_i in S_2 and a surface patch S_p in S_1. Then, compute a numerical value for X_T by solving the constraint in (6.6).
4. Update the corresponding accumulator array by incrementing the votes associated with the estimated parameter in the previous step.
5. Repeat steps 3 and 4 until all plausible point-to-surface patch correspondences have been explored.
6. Select the distinct peak of the accumulator array (Figure 6.13). Update the parameter X_T to the peak value.
7. Repeat steps 2 to 6 until all parameters have been updated.

8. Repeat the above procedure until convergence (i.e., the parameters have not changed more than a predefined threshold between successive iterations); Figure 6.14.
9. Using the associations that contributed to the peak in the last iteration, we can perform a simultaneous least-squares adjustment to come up with an estimate for the transformation parameters relating the two datasets.

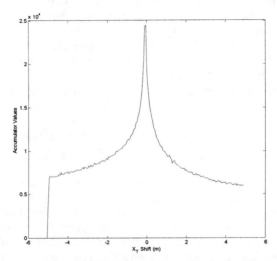

Figure 6.13 Accumulator array for the XT shift of the 3-D similarity transformation.

The above procedure should be implemented in a coarse-to-fine strategy that controls the precision of the solution and the permissible range for the parameter under consideration. Within the iterative procedure, the accumulator array for each parameter is digitized into several cells that span a defined range. In the earlier iterations, the discrete solution steps (cell sizes) are digitized coarsely, to compensate for the poor quality of the approximate values of the unknown parameters. However in later iterations, the solution range as well as the digitization steps (cell size) can be reduced to reflect the improved quality of the updated transformation parameters. At the end of each iteration, the approximations are updated and the cell size of the accumulator array is reduced. In this manner, the parameters would be estimated with high accuracy.

In summary, the above procedure can be used to derive a global similarity measure (σ_o^2) between the involved DTM datasets while establishing the correspondences between conjugate surface elements (points in S_2 and corresponding patches in S_1). In the meantime, the transformation parameters between the associated reference frames are estimated. Note that this procedure can be used to compare temporal datasets for quality control and change detection applications. In such a case, common areas between the two datasets will be used to establish the transformation parameters while incompatible areas, which might be caused by erroneous DTM values and/or changes in the object space, will be filtered out during the iterative voting scheme. After convergence, nonmatched primitives could be highlighted and attributed to errors in the DTM data and/or changes that took place between the moments of acquisition. For a more detailed discussion on this procedure, interested readers can refer to Habib et al. (2001).

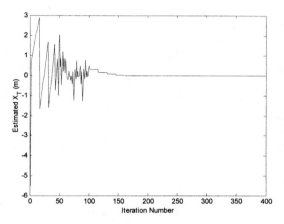

Figure 6.14 XT convergence through the iterative procedure.

References

Bossler, J., Jensen, J., McMaster, R., and Rizos, C. (Eds.), *Manual of Geospatial Science and Technology*, New York: Taylor & Francis, 2002.

Habib, A., Lee, Y., and Morgan, M., "Surface matching and change detection using the modified Hough transform for robust parameter estimation," *Photogrammetric Record*, 17(98): 303-315, Oct. 2001.

Haralick, R., and Shapiro, L., *Computer and Robot Vision: Volume 2*, Reading, MA: Addison-Wesley, 1993.

Kraus, K., *Photogrammetry: Fundamentals and Standard Processes*, Bonn: Dummler, 1993.

Mikhail, M., Bethel, J., and McGlone, J., *Modern Photogrammetry*, New York: John Wiley & Sons, 2001.

Schalkoff, R., *Digital Image Processing and Computer Vision*, New York: John Wiley & Sons, 1989.

Schenk, T., *Digital Photogrammetry: Volume 1*, Laurelville, OH: TerraScience, 1999.

Terei, G., *A Thorough Investigation of Digital Terrain Model Generalization Using Adaptive Filtering*, Ph.D. dissertation, Department of Civil and Environmental Engineering, and Geodetic Science, The Ohio State University, 151, 1999.

Wolf, P., and Dewitt, B., *Elements of Photogrammetry with Applications in GIS*, New York: McGraw-Hill, 2000.

Chapter 7

Mapping and Engineering Applications

Mapping and engineering applications of terrain data often involve some type of spatial analysis and data manipulation. The types of analysis and the degree of manipulation will vary with the application, but the techniques span the range of simple visualization to applications of highly sophisticated mathematical models to the terrain data to derive additional data forms.

Visualization techniques can assist the user in making instantaneous decisions and generally involve little to no data manipulation (Bailey and Gatrell, 1995). Visually *fusing* or combining the terrain data graphically with other datasets can go even further to providing desirable information. Analyses involving slightly more sophisticated manipulations of the data include first- or second-order derivatives of the terrain. Examples include slope and curvature and are the results of applying mathematical equations to create new datasets from the terrain. This chapter will examine some popular visualization techniques for terrain data and some of the more commonly used first- and second-order derivatives of terrain.

7.1 DTM VISUALIZATION

There are a number of basic ways to enhance and display DTM data. These basic techniques can also be combined to create additional display techniques. Other datasets that may be unrelated to the terrain (such as land cover) can be graphically combined or fused with the terrain data to provide information on environmental processes. This latter type of visualization is explored further in Chapter 8.

7.1.1 Grayscale Image

A basic and first step in visualizing DTMs involves assigning a gray value, or brightness value g_i, to each pixel in the DTM based on the elevation Z_i at each grid cell or TIN triangle, and the range of values in the DTM. Equation (7.1) describes the assignment of values where Z_{min} and Z_{max} are the minimum and maximum elevations found in the DTM, respectively.

$$g_i = 255 \left(\frac{Z_i - Z_{min}}{Z_{max} - Z_{min}} \right) \tag{7.1}$$

Figure 7.1(a) shows how the gray values are linearly scaled from 0 to 255.

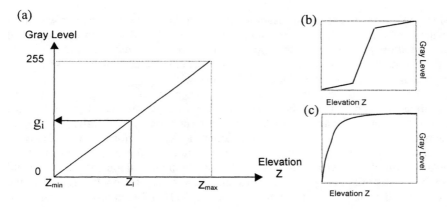

Figure 7.1 Scaling gray values with (a) linear scaling, (b) piece-wise linear function, or (c) gamma function.

In this example, the choice to scale (or "stretch") from 0 to 255 was somewhat arbitrary and was chosen simply due to the constraints of certain monitors and printers that are restricted to 8-bit precision in the gray color. Other radiometric resolutions are possible and the elevations could have been scaled between any two values. If the intention is to color drape a satellite image over a DTM, the choice of scale may be influenced by the radiometric resolution of the sensor. For example, landsat thematic mappers 4 and 5 record data in 8 bits and thus, 256 discernibly different levels of radiance (2^8) are possible. Landsat 1's MSS sensor only recorded data in 6 bits thus producing data that ranged from 0 to 63. Sensors have been developed with resolutions of 12 bits (Jensen, 2000).

Linear scaling is only one possibility, and many other stretching functions exist (Gonzalez and Wood, 2001). Figure 7.1(b) shows how a piecewise linear function can stretch the elevation values between the upper and lower bound. A gamma-type function such as that shown in Figure 7.1(c) is like the piecewise linear function in that it can emphasize values in a particular region of the elevation range. For example, if many of the elevations in the DTM reside in the lower half of the range determined by Z_{min} and Z_{max}, then a gamma function of the type shown in Figure 7.1(c) is appropriate, as it will provide a better stretch to those values in the lower half of the range.

Gray-level shading is also limited by the human visual system (which has a perception of only one to two dozen gray levels). Pseudocolor (or false color) may be a better option as a wider range of discernable shades of color can be detected by the human visual system as well as displayed on screen for many devices.

7.1.2 Shaded Relief

The shaded relief image is a powerful tool that can be used to highlight structure within a DEM. It is also known as "sun" shading because it simulates how a terrain surface would look if the sun were in a specific position in the sky (defined by an azimuth and elevation angle above the horizon). The methodology for producing a shaded relief image is shown by the following:

1. Estimate the normal vector n to the surface, tin element, or grid cell.
2. Choose a light source location vector L, for example, $L = (45°, 45°)$, or alternatively, $L = (1, 1, \sqrt{2})$, both of which are light sources located in the NE corner with an elevation of 45° above the horizontal. A light source located at $L = (0, 0, 1)$ indicates a light source situated at the zenith.
3. Set the gray values of the TIN triangles or grid cells to be proportional to the angle θ between n and L.

Equation (7.2) describes the mathematical assignment of a gray value g to a TIN triangle or grid cell. Figure 7.2 illustrates the application of this concept to both TIN elements and grid cells.

$$g = f(\cos\theta) = 255\left(\frac{n \cdot L}{|n||L|}\right) \quad \text{If } g <= 0, \text{ then } g = 0. \tag{7.2}$$

For a TIN triangle, the vector n is defined to be normal to the triangle surface, and is therefore normal to the two vectors a and b that define the triangle surface. Implicit in Equation (7.2) is the fact that the vectors n and L lie in the same plane that is orthogonal to the triangle surface. If L were to lie in the same plane as the triangle surface, then g would be zero.

In grid models, the vector normal is defined by the vector (a, b, c), which is normal to the surface created by the grid points. The values of a, b, and c are the coordinates in the x-y-z plane of the vector $\boldsymbol{n}$. The grid cell surface (shown shaded) may contain a surface vector with x-y-z coordinates (X, Y, Z) and thus, the vector normal is defined by the expression $aX+bY+cZ = 0$ (Kaplan, 1984).

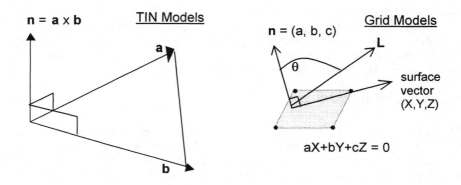

Figure 7.2 Normal vectors in TIN and grid models of terrain.

It is important to notice that this method only shades the surface of the model according to its inclination and orientation. Large objects will not throw shadows onto other areas of the DTM. The images shown in Figure 7.3 illustrate the shaded relief concept. Figure 7.3(a) provides a grayscale image of a region crossing southern Alberta and southern British Columbia, Canada. The latitude and longitude coordinates are shown in the figure. The shaded relief map provides a representation of the DTM with vector L as (325°, 45°) and Figure 7.3 (c) shows the visualization of this same terrain by draping the grayscale image over the shaded relief map (with some additional scaling of height). Shaded relief layers are often set up as intensity layers, so that they can be combined with pseudocolor layers to generate color-drape images. Figure 7.3 provides an almost 3-D rendering of the digital terrain model.

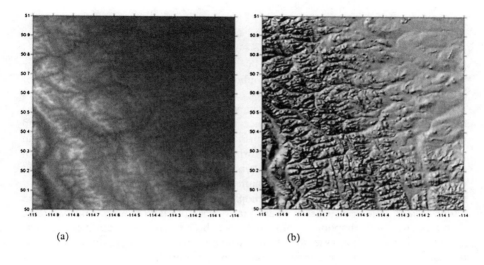

(a)   (b)

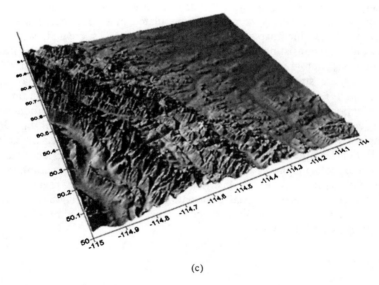

(c)

Figure 7.3 Visualizing southern Albertan and British Columbian terrain. (a) Grayscale image (higher elevations are brighter), (b) shaded relief image, and (c) composite of (b) draped over (a). Terrain data courtesy of GeoBase.

7.2 DTMS AND ORTHOPHOTOS

In the same way that a grayscale image was draped over a shaded relief map to provide a 3-D rendering of the DTM in Figure 7.3, combining DTMs with other types of information, particularly orthophotos, can provide information on line of sight, locations of hydroelectric projects, and so forth.

Orthophoto maps are relatively inexpensive to produce and provide more data than conventional line maps. Since they do not require as much time and effort to produce as traditional maps, orthophotos represent a good way to obtain and maintain up-to-date geographic information. Orthophoto maps do not contain many of the nonvisible features and attributes, such as political boundaries, street names and route numbers, and place names and address ranges that often appear on map documents. Nevertheless, orthophotography is a good foundation upon which to assemble and represent these kinds of spatial data from various sources. Applications for orthophotography include a broad range of activities such as parcel mapping and tax assessment, infrastructure management, resource analysis, drainage studies, and emergency 911 operations mapping.

Many orthomapping projects today involve the creation of digital orthoimagery that can be displayed and manipulated in a computer. Instead of a hard copy photographic map document, a *digital ortho* is created, which is a computer data file consisting of a collection of numbers. Each number represents the brightness (photographic gray tone) of a small patch of the ground. The numbers for these small patches are displayed on a computer screen as square cells called *pixels*. The ground size of this patch (or pixel) is the resolution of the digital orthophoto. Digital orthophotography today is typically produced by scanning conventional film-based aerial photographs. The resulting digital photographs are then transformed in a computer into an *orthorectified* image map. It is likely that satellite sources will become an increasingly important source of data for orthophoto mapping. It is important to note that digital orthophotos are not GIS databases; they are collections of pixels that may require further processing to obtain useful spatial information to support decision-making. For example, extracting the road network centerlines from the orthoimage can be time-consuming, difficult work.

7.2.1 Orthoimages

An orthophoto or orthoimage is an aerial photograph or image that has the geometric properties of a map. Thus, orthoimages can be used as maps to provide measurements and establish accurate geographic locations of features. Orthoimages generated from aerial photographs and satellite images have undergone the process known as *orthorectification*. An unrectified aerial photograph or satellite image will not show features in their correct locations due to displacements caused by the tilt of the sensor and by the relief in terrain. Orthorectification transforms the central projection of the image into an orthogonal view of the ground, thereby removing the distorting effects of tilt and terrain relief, as shown in Figure 7.4.

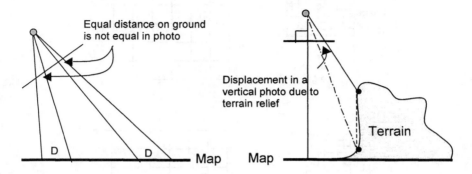

Figure 7.4 Displacement in aerial photos due to sensor tilt and terrain relief.

In this figure, the tilt of the camera causes two equal distances on the ground to be perceived as unequal in the photo. In addition, two points that should correspond to the same horizontal location in a map will appear in different locations in an aerial photo due to their differences in elevation.

Subsequently, generation of an orthophoto map from an aerial photograph requires information on the location of the camera and its orientation in space, as well as a model of the terrain elevation. If the terrain is very flat, then ground elevation data may not be required. In such situations, orthophoto maps can be produced by a process called simple (perspective) rectification that only removes the effect of tilt. However in most cases, it may be necessary to perform the more involved orthorectification process to obtain an accurate map.

7.2.2 Rectification of Orthophotos

Rectification is the process of allocating a gray value from the image to a location on the DTM. There are generally two types of rectification: *direct* and *indirect,* and they are illustrated in Figure 7.5.

In the indirect method, the user is conducting a transformation from the DTM coordinates (X, Y) to the image coordinates (x, y), while in the direct method, the user is conducting a transformation from the image coordinates (x,y) to the DTM coordinates (X, Y). Table 7.1 describes the general methodology and the advantages and disadvantages to each method.

A description of two rectification techniques is provided in Section 7.2.3. They are each suited to different cases depending on the type of data used.

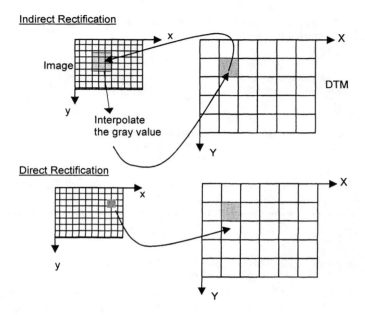

Figure 7.5 Difference between indirect and direct rectification.

Table 7.1

Comparison Between Direct and Indirect Rectification

	Rectification Techniques	
	Indirect	*Direct*
Steps	1. Transform from the DTM coordinates (X, Y) to the image coordinates (x,y). 2. Estimate the gray/color value using interpolation techniques (e.g., nearest neighbor). 3. Assign the interpolated gray/color value to the transformed DTM cell.	1. Transform from the image coordinates (x,y) to the DTM coordinates (X,Y). 2. Assign the pixel gray/color value to the nearest DTM cell.
Advantages	➤ Every DTM cell will get a gray/color value.	➤ Gray/color values of the image do not change.
Disadvantages	➤ Interpolating the gray/color value is time-consuming. ➤ The gray/color values of the final rectified images are not the same as those of the original image (due to interpolation).	➤ Not all the DTM cells will be assigned a gray/color value from the image and therefore, their gray/color values have to be interpolated from neighboring cells.

7.2.3 Polynomial Rectification

Polynomial rectification may be applied as a direct or indirect rectification. It uses ground control points (GCPs) to relate the DTM coordinate system to the image coordinate system. The degree of the polynomial will depend on the number of GCPs used and the more GCPs that are used, the more accurate the results of the rectification will be.

Polynomial rectification is completely independent of the geometry of the image, and therefore can be used for both satellite and aerial images. However, it is more often used for satellite images due to the following reasons:

1. Image geometry and distortions are sometimes difficult to model.
2. The relief displacement due to the topography of the Earth is relatively small compared to the flying height of the satellite and does not influence the result significantly.

These reasons are true because polynomial rectification does not correct for terrain displacement. The general formula for the rectification is shown in (7.3), while (7.4) shows the formula when $N=2$ (i.e., second-order polynomial).

$$x = \sum_{i=0}^{N} \sum_{j=0}^{N-i} a_{ij} X^i Y^j \tag{7.3}$$

$$y = \sum_{i=0}^{N} \sum_{j=0}^{N-i} b_{ij} X^i Y^j$$

$$x = a_{00} + a_{10}X + a_{01}Y + a_{20}X^2 + a_{11}XY + a_{02}Y^2$$
$$y = b_{00} + b_{10}X + b_{01}Y + b_{20}X^2 + b_{11}XY + b_{02}Y^2 \tag{7.4}$$

In (7.3) and (7.4), (X, Y) are the coordinates of each cell in the DTM, (x, y) are the coordinates in the corresponding cell in the image, and N is the order of the polynomial. Table 7.2 illustrates the role of each coefficient shown in (7.4).

Table 7.2

Coefficients in (7.4)

Coefficient	Warp Component
a_{00}; b_{00}	Shift in x; Shift in y
a_{10}; b_{10}	Scale in x; Scale in y
a_{01}; b_{01}	Shear in x; Shear in y
a_{20}; b_{20}	Nonlinear scale in x; nonlinear scale in y
a_{11}; b_{11}	y-dependant scale in x; x-dependent scale in y
a_{02}; b_{02}	Nonlinear shear in x; nonlinear shear in y

The advantages to polynomial rectification include that the method is easy to implement and all the distortions in the image are due to sensor geometry, earth curvature, and so forth, and are corrected simultaneously. The disadvantages include that the method does not correct for relief displacement and the method does not consider the geometric model of the imaging system (e.g., collinearity).

7.2.4 Differential Rectification

This method requires: (1) an image and a DTM covering the same area of the image; (2) the exterior orientation parameters of the image; that is the position vector (three parameters per image) and orientation (three parameters per image) of the camera at exposure times; and (3) the interior geometry of the camera (lens focal length and the lens distortion parameters) as estimated from the camera calibration). The objective of differential rectification is the assignment of gray values from the image to each cell of the DTM. After the rectification, both the elevation and the gray/color values are stored at the same location of the DTM.

To determine the gray/color values, the 3-D coordinates (X_i, Y_i, Z_i) of each DTM cell are transformed into the image domain using the collinearity equations shown in (7.5) and (7.6). The process is illustrated in Figure 7.6.

As the equations show, the image position (x_i, y_i) that corresponds to the DTM location (X_i, Y_i, Z_i) by collinearity is determined using the coordinates of the camera's perspective center (X_o, Y_o, Z_o) (see arrow 1 in Figure 7.6). The gray/color value for the cell (x_i, y_i) is interpolated by one of the resampling methods (e.g., nearest neighbor). This gray/coordinate value is then assigned to the DTM cell at the (X_i, Y_i) location (see arrow 2 in Figure 7.6).

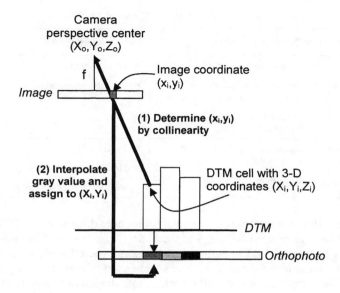

Figure 7.6 Digital rectification of an orthophoto.

The collinearity equations (describing the spatial relationships among image coordinates, perspective centers, and object-space coordinates) are shown next.

$$x_i - x_p = -f \frac{m_{11}(X_i - X_0) + m_{12}(Y_i - Y_0) + m_{13}(Z_i - Z_0)}{m_{31}(X_i - X_0) + m_{32}(Y_i - Y_0) + m_{33}(Z_i - Z_0)} \tag{7.5}$$

$$y_i - y_p = -f \frac{m_{21}(X_i - X_0) + m_{22}(Y_i - Y_0) + m_{23}(Z_i - Z_0)}{m_{31}(X_i - X_0) + m_{32}(Y_i - Y_0) + m_{33}(Z_i - Z_0)} \tag{7.6}$$

where (x_i, y_i) are image coordinates of point i; x_p, y_p are the coordinates of the principal point expressed in the image coordinate system; (X_i, Y_i, Z_i) are the 3-D object-space coordinates of the DTM cell; (X_0, Y_0, Z_0), are the 3-D coordinates of the perspective center; and m_{ij} are the coefficients of the rotation matrix. Table 7.3 compares each method and describes the advantages and disadvantages to each.

Table 7.3

Comparison Between Polynomial and Differential Rectification

	Polynomial Rectification	*Differential Rectification*
Advantages	➤ Fast and easy to create ➤ Can use cheap scanner (for scanning aerial images) ➤ Can use map to determine the GCP coordinates	➤ More accurate ➤ Uses the sensor model
Disadvantages	➤ Less accurate (does not use the sensor model) ➤ Needs more GCPs	➤ Requires DTM ➤ Needs expensive scanner ➤ Requires the sensor model

Figure 7.7 shows the draping of a digital orthophoto draped over a DTM. The figure shows several campus buildings at the Federal University of Paraná–UFPR in Brazil. The background DTM is shown in grayscale where the higher elevations are brighter than the lower elevations. The square shown in the lower right is the depiction of the completely orthorectified orthophoto draped over the DTM.

Figure 7.7 Orthophoto draped over a DTM of the Federal University of Paraná–UFPR in Brazil.

7.3 FIRST-ORDER DTM DERIVATIVES

DTM *derivatives* are simply data products derived from DTM data. They often involve *neighborhood operations* whereby analysis of a "neighborhood" or group of adjacent cells in a DTM is conducted to produce the new data. The term derivative also implies that the operation is based on the change or difference in elevation between one cell in the DTM and those within this neighborhood. For example, slope maps are maps of the rate of change, or the first derivative of a DTM (Gallant and Wilson, 2000). Here, the neighborhood operation is performed on the DTM in order to get a measure of the steepness of an area of the Earth's surface.

7.3.1 Slope

Slope is usually measured in degrees or percent and is defined as the rate of change in elevation Δz over a change in lateral extent. The rate of change of elevation in both the x and y directions can be used to identify the direction and magnitude of the steepest gradient. These two parameters can be found by taking the partial first-order derivatives of the elevation z with respect to x and y. Therefore, slope (magnitude) can be found by combining the two component partial derivatives, as shown in (7.7):

$$\text{Slope} = \frac{\delta Z}{\delta XY} = \sqrt{\left(\frac{\delta Z}{\delta X}\right)^2 + \left(\frac{\delta Z}{\delta Y}\right)^2} \tag{7.7}$$

Elevation will change in both the x and y directions, and hence the slope resulting from a change in lateral extent is the vector sum of the slope in the x direction and the slope in the y direction. Figure 7.8 illustrates these concepts.

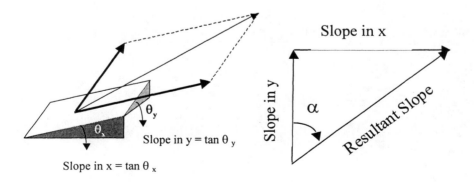

Figure 7.8 Definition of slope.

Note that

$$\text{Slope} = \sqrt{\left(\frac{\delta f}{\delta X}\right)^2 + \left(\frac{\delta f}{\delta Y}\right)^2} = \sqrt{\left(\frac{\Delta Z_x}{\Delta X}\right)^2 + \left(\frac{\Delta Z_y}{\Delta Y}\right)^2} \tag{7.8}$$

and the angle of the slope with respect to the horizontal is equal to the tangent of the slope, and α is the aspect of the slope (see Section 7.3.2).

With regard to grid data, the resulting slope value assigned to each cell will reflect the overall slope based on the relationship between that cell and its neighbors. There are numerous ways that slope can be calculated when dealing with grid data. The illustration in Figure 7.9 shows the use of four neighboring points.

Z(1,1)	Z(1,2)	Z(1,3)
Z(2,1)	Z(2,2)	Z(2,3)
Z(3,1)	Z(3,2)	Z(3,3)

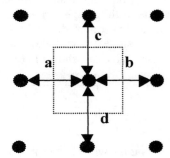

Figure 7.9 Slope derived from four neighbors.

Thus, Figure 7.9 implies that

$$\begin{aligned}
\frac{\delta f}{\delta X} &= \frac{\text{Slope(a)} + \text{Slope(b)}}{2} \\
&= \frac{1}{2}\left[\frac{Z(2,2) - Z(2,1)}{D} + \frac{Z(2,3) - Z(2,2)}{D}\right] \\
&= \frac{Z(2,3) - Z(2,1)}{2D}
\end{aligned} \tag{7.9}$$

$$\frac{\delta f}{\delta Y} = \frac{\text{Slope(c)} + \text{Slope(d)}}{2}$$

$$= \frac{1}{2}\left[\frac{Z(2,2) - Z(1,2)}{D} + \frac{Z(3,2) - Z(2,2)}{D}\right] \qquad (7.10)$$

$$= \frac{Z(3,2) - Z(1,2)}{2D}$$

where D is the resolution of the grid cells. Figure 7.10 illustrates the computation of slope using eight neighbors. In this situation, the corresponding equations are:

$$\frac{\delta f}{\delta Y} = \frac{\sum\limits_{i=1}^{6}\text{Slope(S}_i)}{6} = \qquad (7.11)$$

$$\left[\frac{Z(3,1) + Z(3,2) + Z(3,3) - Z(1,1) - Z(1,2) - Z(1,3)}{6D}\right]$$

$$\frac{\delta f}{\delta Y} = \frac{\sum\limits_{i=1}^{6}\text{Slope(S}_i)}{6} = \qquad (7.12)$$

$$\left[\frac{Z(3,1) + Z(3,2) + Z(3,3) - Z(1,1) - Z(1,2) - Z(1,3)}{6D}\right]$$

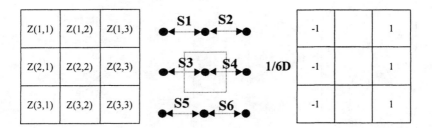

Figure 7.10 Deriving slope using the eight neighbors.

Figure 7.11 shows a portion of the National Topographic Survey's map sheet 82O encompassing Cochrane, Alberta. Figure 7.12 shows the grayscale image of the DTM and Figure 7.13 shows the slope map derived for this region.

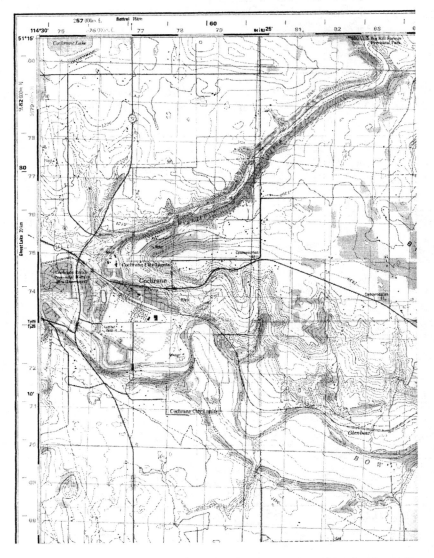

Figure 7.11 Portion of NTS Map Sheet 82O/1 showing Cochrane, Alberta. (© Produced under license from her Majesty the Queen in Right of Canada, with permission of Natural Resources Canada.)

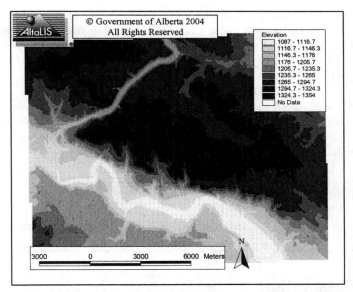

Figure 7.12 Grayscale image depicting the DTM (elevation in meters) for the town of Cochrane, Alberta and the surrounding region. (Data courtesy of AltaLIS and all rights are reserved by the Government of Alberta, 2004.)

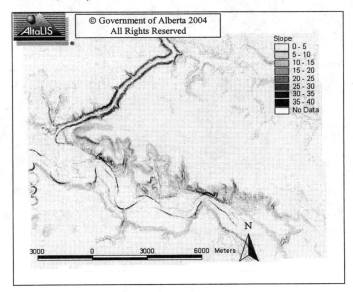

Figure 7.13 Slope map (in degrees) derived from the DTM given in Figure 7.11 using the eight neighbor approach. (Data courtesy of AltaLIS and all rights are reserved by the Government of Alberta, 2004.)

The Bow River and Big Hill Creek (to the north) are clearly shown in Figure 7.12, but Figure 7.13 shows that the terrain surrounding Big Hill Creek is much steeper than the terrain on either side of the Bow River.

7.3.2 Aspect

The aspect value assigned to each cell in an aspect map tells us the direction (north, south, and so forth) to which that cell is oriented. Equation (7.13) describes how that direction is computed where α is computed as the angle clockwise from north (see Figure 7.8).

$$\tan \alpha = \frac{\delta f}{\delta X} / \frac{\delta f}{\delta Y} \qquad (7.13)$$

An aspect map of the DTM shown in Figure 7.12 is provided in Figure 7.14.

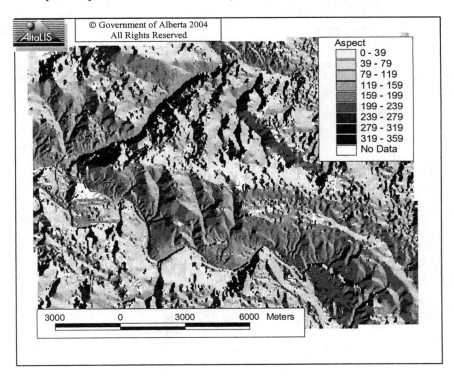

Figure 7.14 Aspect map (in degrees) for the DTM given in Figure 7.12. (Data courtesy of AltaLIS and all rights are reserved by the Government of Alberta, 2004.)

Notice in the figure above that the valley defined by Big Hill Creek faces northwest while the region on the southern side of the Bow River faces north.

7.4 SECOND-ORDER DTM DERIVATIVES

The second derivative of a DTM (the first derivative of the slope) describes the slope rate of change of the terrain. It also describes the curvature of the terrain in terms of how convex, concave, or straight each grid cell is configured. Curvature maps provide a measure of the rate of change of slopes. Geomorphologists use curvature maps in landform curvature analysis (convex/concave) and aging of terrain studies (change detection). Other applications include flow acceleration, soil water property analysis, and land evaluation. Some of these applications are further described in Chapter 8. Figure 7.15 describes how curvature is computed.

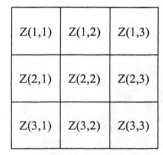 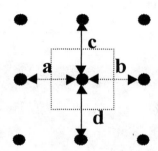

Figure 7.15 Deriving curvature from a grid.

The resulting equations for deriving curvature are:

$$\text{Curvature} = \sqrt{\left(\frac{\delta^2 f}{\delta X^2}\right)^2 + \left(\frac{\delta^2 f}{\delta Y^2}\right)^2} \qquad (7.14)$$

where

$$
\begin{aligned}
\frac{\delta^2 f}{\delta X^2} &= \frac{\text{Slope(b)} - \text{Slope(a)}}{D} \\
&= \frac{1}{D}\left[\frac{Z(2,2) - Z(2,1)}{D} - \frac{Z(2,3) - Z(2,2)}{D}\right] \\
&= \frac{2Z(2,2) - Z(2,3) - Z(2,1)}{D^2}
\end{aligned}
\qquad (7.15)
$$

$$\frac{\delta^2 f}{\delta Y^2} = \frac{\text{Slope}(d) - \text{Slope}(c)}{D}$$

$$= \frac{1}{D}\left[\frac{Z(2,2) - Z(1,2)}{D} + \frac{Z(3,2) - Z(2,2)}{D}\right] \qquad (7.16)$$

$$= \frac{2Z(2,2) - Z(3,2) - Z(1,2)}{D^2}$$

Figure 7.16 depicts a curvature map for the DTM shown in Figure 7.11. Negative curvature values indicate that the terrain is upwardly concave at the cell. Values of zero curvature are due to flat cells and positive values of curvature indicate that the surface is upwardly convex at the cell. Notice that in Figure 7.16, many of the cells show relatively flat curvature but the regions near streams show areas of convex terrain (usually pertaining to mountainous areas) developing into convex curvature (valley-type formations) as you move closer to the center of the river.

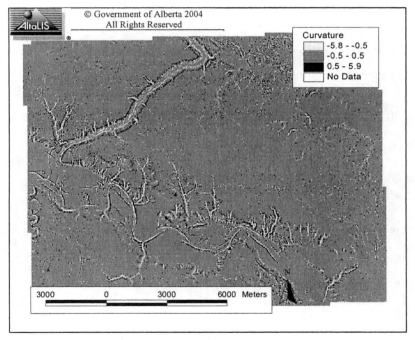

Figure 7.16 Curvature map (units of 0.01 m^{-1}) for the DTM given in Figure 7.12. (Data courtesy of AltaLIS and all rights are reserved by the Government of Alberta, 2004.)

7.5 VOLUME COMPUTATIONS

7.5.1 Volume Computations from Contour Data

Very often computations of earth volume are essential parameters in transportation planning, construction, and other civil and military applications. The principle involves splitting the ground along the contour planes into a series of horizontal slabs. Each slab being considered is a prismoid with the height equaling the contour interval and end areas are the areas enclosed by the contour lines. The volume of the prismoid V_{i-j}^c between the two contours i and j is equal to:

$$V_{i-j}^c = CI \times \left[\frac{A_i^c + A_j^c}{2} \right] \tag{7.17}$$

or more generally,

$$V_{i-j}^c = \frac{CI}{2} \times \left[A_i^c + 2(A_{i+1}^c + A_{i+2}^c + \ldots\ldots A_{j-2}^c + A_{j-1}^c) + A_j^c \right] \tag{7.18}$$

where CI is the contour interval and A^c_i is the area contained by contour i.

The accuracy of the volume computation is a function of the accuracy of the contour map, the contour interval, and the accuracy in measuring the area enclosed by the contour. Consider the following example illustrated in Figure 7.17. The problem is to calculate the volume of cut required to level the area shown in the figure to 5.0m.

$$\text{Volume} = V_{5-25}^c + V_{25-\text{point}28}^c = \frac{5}{2} \times \left[A_5^c + 2(A_{10}^c + A_{15}^c + A_{20}^c) + A_{25}^c \right] + \frac{28-25}{2} \times A_{25}^c$$

Note that the volume between contour 25 and point 28 is equal to:

$$V_{25-\text{point}28}^c = \frac{\text{Height}}{2} \times A_{\text{base}}$$

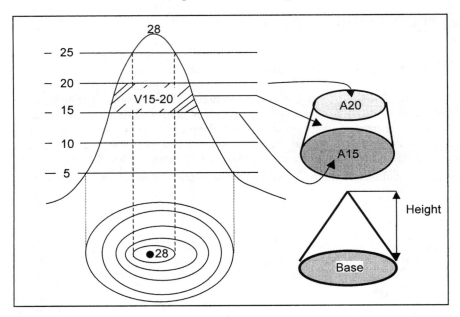

Figure 7.17 The volume derivation.

7.5.2 Volume Computations from Grid Data

The principle for volume computations from grid data involves the volume within each DTM cell which is taken as being:

$$\text{Volume} = A_{\text{cell}} \times H_{\text{average}}$$
$$H_{\text{average}} = \frac{H_1 + H_2 + H_3 + H_4}{4} \tag{7.19}$$

where H_i is the depth of cut or fill at the corners of the cell ($i = 1 - 4$) and the area of each cell is $A_{\text{cell}} = D^2$ for grids of resolution D.

or generally,

$$\text{Volume} = \frac{D^2}{4}\left[\sum H_1 + 2\sum H_2 + 3\sum H_3 + 4\sum H_4\right] \tag{7.20}$$

where H_1 is a depth being used once in the computation; H_2 is a depth being used twice in the computation; H_3 is a depth used three times in the calculation, and H_4 is a value being used four times. In Figure 7.18, the shaded region encompasses all those grid points with depth values that are used four times in the computation. The other grid points are labeled according to the number of times the depth values are used in the computation.

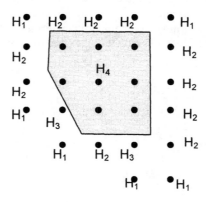

Figure 7.18 Illustration of volume computation using grid data.

7.5.3 Generating Contour Lines from Grid Data

If contour data is unavailable and the only data is a raster grid, contour lines can be generated from the grid data. Most commercial contour software developed for this purpose use *contour isolines* to delineate lines across grid cells. The following method is an automatic procedure that allows a user to extract contour lines systematically. It starts by conducting the following initial steps:

1. Estimate the Zmin and Zmax in the grid area.
2. Decide on the contour interval (CI).
3. Calculate the first and last contour line elevations using:
   - First contour line elevation = [truncate (Zmin) + CI];
   - Last contour line elevation = [truncate (Zmax)].

Figure 7.19 illustrates some of the next steps used in generating the contour lines. Starting with the first contour line, step through the following procedure:

1. Loop through the grid cell by cell. If all the elevations of the four cell nodes are less than or greater than the contour level (i.e., the contour height), then

there will be no contour line passing through this cell. If the condition in step
(1) is not satisfied then it means that the specified contour (in this case the
first contour) will pass through this cell.

2. To find the contour points in the cell, start at the first pair of nodes of the cell
    and determine if the contour exists along this edge. This occurs when one
    node height is less than the contour level while the other node is greater than
    the contour level. If no contour point exists along this edge, proceed in either
    a consistent clockwise or counterclockwise direction until an edge containing
    the contour point is found.

3. Once an edge containing a contour point is found between the grid nodes i
    and j, compute the location of the contour point by linear interpolation, as
    shown in (7.21):

$$\frac{Z_c - Z_i}{Z_j - Z_i} = \frac{X_c - X_i}{X_j - X_i}$$

$$X_c = X_c + \frac{Z_c - Z_i}{Z_j - Z_i} \times (X_j - X_i)$$

$$Y_c = Y_c + \frac{Z_c - Z_i}{Z_j - Z_i} \times (Y_j - Y_i)$$

(7.21)

where (X_c, Y_c, Z_c) are the coordinates of the contour point (and therefore, Z_c is
the elevation of that contour line), and (X_i, Y_i, Z_i) and (X_j, Y_j, Z_j) are the
coordinates of the grid nodes i and j confining the contour point.

4. Examine each subsequent edge until the next edge containing a contour point
    is found. Repeat step 3 and connect these two pints to form the contour line
    and store the coordinates of the contour points as these coordinates will be
    used in calculating the area enclosed by the contour line for volume
    computations.

5. Follow the azimuth of the contour line to move to neighboring grid cells.
    Repeat steps 2–4 until you either close the contour line or the contour exits
    the grid.

6. Increment CI and repeat steps 2–5 until you reach the last contour.

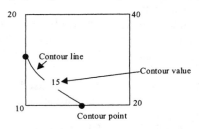

Figure 7.19 Illustration of step (1) in determining contour lines from grid data.

7.6 LINE-OF-SIGHT APPLICATIONS

A variety of applications that require *line-of-sight* information use high-resolution DTMs. Line-of-sight, or the path of visibility for an observer, is used extensively in urban planning (parkland siting), military planning (border patrols, security camera locations, GNSS signal availability), and wireless civilian applications (determining the most appropriate locations for cell-phone towers and radio transmitters).

The computation of line-of-sight involves asking if a distant point is visible to the observer. Often in line-of-sight applications, the user is scouting potential locations for a facility and is interested in knowing all the points that are visible from the facility's potential location. For this type of application, DTMs are used to create *viewsheds* that map out all areas visible to the potential observer location (ESRI, 1998). Creating a viewshed is not a simple process and can be computationally intensive. It requires determining the relationship between the observer and every point in the DTM, and thus, grid cell sizes that are too fine may not be recommended. However, if the grid resolution is too coarse, it can lead to inaccuracies because peaks and valleys will be misrepresented.

The first step is to assign a *local horizon* to the observer. This is also known as the *visible horizon* and will determine the extent that is visible to the observer and unobstructed by higher terrain. The local horizon is often set to be perfectly horizontal but depending on the flexibility of the algorithm or software being used to compute the viewshed, the local horizon can actually be set to an angle elevated above the horizontal. The visibility of each point (TIN center or grid cell center) in the DTM is determined by comparing the elevation of the observer to the point of interest. If the point lies above the local horizon but unobstructed by higher terrain between the observer and the point, then it is considered to be visible. Note that some algorithms also prevent the observer from looking below the local horizon (in very strict cases), but in many cases, objects (cells) below the local horizon are also visible so long as they are above the slope line made between the observer and any intervening cells. This is illustrated in Figure 7.20.

Note that if the distances being considered are very large across the DTM, the user may need to correct for curvature effects and for refraction. In Figure 7.20, the grid cells that are shaded in gray are visible to the observer but the grid cells shown in white are not, even though their elevations may be higher than the observer. The two cells shown as dotted may or may not be visible depending on the flexibility of the algorithm. In a very strict algorithm that does not allow the viewer to see cells below the local horizon, those two cells would not be visible. In more flexible algorithms that allow the viewer to see below the local horizon, those two dotted cells would be visible, but the third and fourth (from the right) cells would continue to be invisible to the observer because the slope angle between the first and second cells is not steep enough to allow the viewer to see the third and fourth cells.

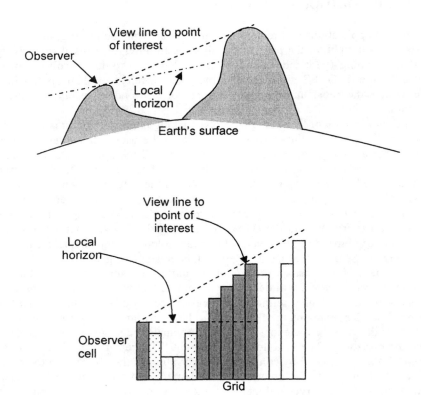

Figure 7.20 Illustration of the viewshed concept.

Figure 7.21 shows areas seen by an observer in the lower right indicated as a star. The visible areas are shown as polygons in black outline. The DTM is shown as a digital grayscale in the background and is the same data used to create the images in Figure 7.3.

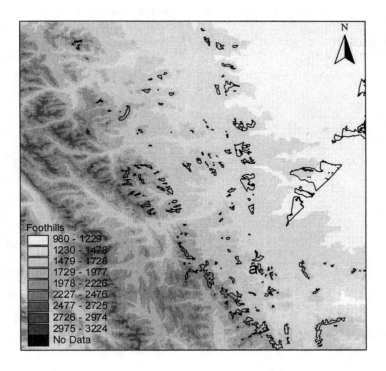

Figure 7.21 Viewshed map of Figure 7.3 with elevation values shown in grayscale in the background (in meters). The black polygons are the areas visible by the observer located at the circled star in the lower right. (Terrain data courtesy of GeoBase).

Not a great deal would seem visible to the observer but that is because the observer is located in a very small localized depression immediately surrounded by mostly higher terrain. Thus, all that is visible is that terrain which is higher than its immediate neighborhood.

References

Bailey, T.C., and Gatrell, A.C., *Interactive Spatial Data Analysis*, Essex, United Kingdom: Pearson Education Limited, 1995.

ESRI, ArcView 3.2 Help Documents, Redlands, CA: ESRI, 1998.

Gallant, J.C., and Wilson, J.P., "Primary topographic attributes," *Terrain Analysis: Principles and Applications*, Wilson J.P., and Gallant, J.C., (Eds.), New York: John Wiley & Sons, 479, 2000.

Gonzalez, R.C., and Wood, R.E., *Digital Image Processing*, 2nd ed., Upper Saddle River, NJ: Prentice Hall, 2001.

Jensen, R., *Remote Sensing of the Environment: An Earth Resource Perspective*, Upper Saddle River, NJ: Prentice Hall, 2000.

Kaplan, W., *Advanced Calculus,* 3rd ed., Reading, MA: Addison-Wesley, 1984.

Chapter 8

Applications in Environmental Modeling

A model of terrain is often a necessary requirement in identifying, analyzing, and mitigating problems in many fields including hydrology, geomorphology, and environmental modeling. In this chapter, we provide some illustrations of the ways in which digital terrain models can be used in analysis of the environment. Note that this chapter is by no means intended to detail the full range of applications of DTMs to environmental modeling. Instead, this chapter will highlight the more common uses for DTMs and their impact on modeling results.

Table 8.1 provides some insight into the range of possible uses of DTMs in environmental modeling. Huggett (2003) noted that both *primary* attributes (first-order derivatives, or parameters derived directly from elevation) and *secondary* attributes (second-order derivatives, or parameters stemming from first-order derivatives) are used extensively in geomorphology and environmental modeling. The table includes some attributes described by Huggett (2003).

As Table 8.1 demonstrates, DTMs are used throughout all spheres of the Earth's system, including the atmosphere (climatic variables), the hydrosphere (overland and subsurface flow), and the terrasphere (vegetation patterns and erosion). The derivation of several of these attributes will be described in this chapter and will begin with one of the most fundamental uses of DTMs in the environment: drainage characterization.

8.1 DRAINAGE MODELING

Water resources engineering is a distinct branch of environmental modeling that deals primarily with water quantity and quality management. Because of its importance to society, sustainable water resources engineering has evolved into its own discipline separate from traditional environmental engineering. Whether mitigating floods, assessing reservoir water supplies, or making decisions about wastewater treatment and disposal, water resources engineering is a large-scale activity that is impacted by terrain at all levels.

Table 8.1

Primary Attributes Derived from DTMs and Their Applications (Adapted from Huggett, 2003)

Attribute	Application
Altitude	Climatic variables (temperature, pressure, and so forth); Vegetation and soil patterns; Potential energy determination.
Slope	Overland and subsurface flow; Erosion.
Aspect	Solar irradiance; Evapotranspiration.
Curvature	Flow acceleration; Erosion and deposition patterns and rates; Soil and land evaluation indices.
Drainage basin slope, area, and length	Time of concentration; Contributing volumes to runoff; Runoff attenuation times and velocities.
Stream lengths and slopes	Channel flow rates and velocities; Erosion rates and sediment yields.

Water resources engineers require advanced knowledge of hydrological science and hydrological modeling. Hydrology is the science of the movement and distribution of the Earth's waters through all spheres of the Earth's system. To characterize and model the distribution of water level (whether in a river or as soil moisture in the ground) in a region is a fundamental requirement of water resources management. Consequently, one of the most basic modeling units within hydrological modeling is the *watershed*. Alternatively known as the *catchment* or *basin*, a watershed is defined as a control volume such that all water entering the control volume drains to one specific point. This drainage point is also referred to as the watershed *outlet* and is the most downstream point in the drainage area. All water falling within the boundary of this watershed will drain through the outlet.

Figure 8.1 illustrates a large watershed of 1,200 km^2 in southern Alberta known as the Elbow River Watershed. The river network depicted drains into the Glenmore Reservoir, which feeds part of the city of Calgary's water supply. The change in elevation from the headwater (furthest most upstream point in the watershed) to the most downstream point is almost 1,000m.

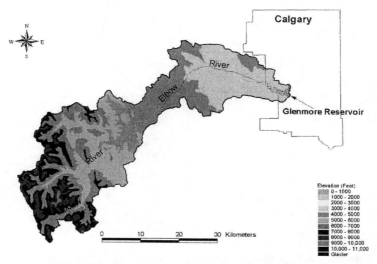

Figure 8.1 The Elbow River Watershed draining into the Glenmore Reservoir near Calgary, Alberta. (Courtesy of Alberta Environment.)

The determination of the watershed boundary and how water accumulates within the watershed are easily determined from DTMs, just as it was for this watershed. But before detailing exactly how DTMs are used to determine watershed boundaries and flow accumulation, Section 8.1.1 provides some terminology and issues dealing with drainage.

8.1.1 Surface Water Drainage and Watersheds

Water poured over a hard perfectly smooth surface will run off or drain by gravity in the direction of steepest descent (when disregarding inertial effects). Routes of steepest descent are called *slopelines* or *flowlines*, and these lines are perpendicular to contour lines (which are lines of constant elevation). This runoff process leads to the development of surface *drainage networks*, which result from the accumulation of flow from merging flowlines. The location and nature of these drainage networks, or river networks, ultimately lead to the definition and delineation of a watershed.

Accurate hydrological modeling requires the accurate representation of a river network. River networks are often idealized as *dendritic* systems that look like trees, where a single branch (or flowline) flows into a larger branch, which flows downstream to yet another larger branch. This is indicated in the illustration on the left in Figure 8.2.

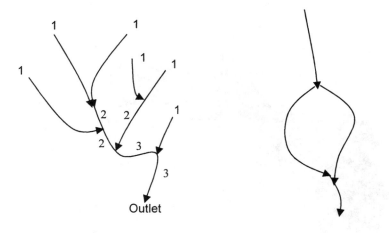

Figure 8.2 Representation of stream networks as dendritic (left) or anabranching (right).

The diagram on the left illustrates this dendritic quality where single river channels each meet at a confluence with a larger channel. This is unlike the illustration on the right in which a river has split off into two channels, which are then rejoined into one channel. These exist in reality and are termed by geomorphologists as *anabranching* rivers. They present a problem when attempting to develop river networks using DTMs because the derivation of river networks by DTMs almost exclusively relies on the assumption of a dendritic network. One other useful quality of stream networks seen in Section 8.4 is the concept of a *stream order*. Horton (1932) and Strahler (1957) are considered pioneers in the quantification of river networks. A stream of lowest stream order 1 is a stream with no other tributaries (smaller rivers) draining into it. As streams converge at confluence points, streams of higher order are created. The stream order is assigned such that when two streams of the same order converge, then the order of the new stream is the old stream order number plus 1. However, if two streams converge with different stream order numbers then the new stream is assigned the higher of the two stream order numbers (there is no increase in stream order number from the two original stream order numbers). Similarly, catchments are also assigned orders according to the Horton's law of basins, which implies that the drainage area of a catchment is directly proportional to the stream order of that catchment. This becomes an important issue when determining stream networks from DTMs.

Consequently, DTMs represent an ideal mechanism by which to model drainage patterns. However, before defining a drainage network or a watershed boundary, the user must first resolve the issue of pits and sinks, as these are phenomena that obstruct the accurate determination of drainage networks and boundaries.

8.1.2 Sinks and Peaks

Errors in DTMs, particularly when generated from contour lines and spot data, can manifest themselves as artificial topographic features such as *sinks* (or depressions) and *peaks*. A sink is a cell value with a lower elevation than all of its surrounding eight neighbors while a peak has a higher elevation than all of its eight neighbors. Figure 8.3 illustrates the result of a peak and a sink on a surface created using KRIGING. Both of these features represent problems when developing stream networks because as indicated earlier, stream networks are generally derived under the assumption of being dendritic, and that all areas in a watershed are "hydrologically connected" (all cells contribute flow to the outlet). The outlet is the most downstream point in the watershed, and hence no internal sinks can exist in the DTM as water flowing into the sink would not be allowed to flow to the outlet. The boundary of a watershed is the area's drainage divide, and hence contains the highest elevations within the watershed. Therefore, peaks represent internal drainage divides that contradict the definition of a watershed. Peaks and sinks are phenomena that need to be detected and removed before developing drainage networks.

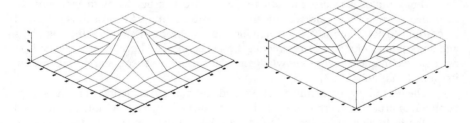

Figure 8.3 A 3-D rendering of a peak (left) and a sink (right), in a surface that was developed using KRIGING.

Sinks and peaks are caused by data inaccuracies and errors (Raaflaub, 2002). According to Raaflaub (2002), there are a variety of techniques to perform sink removal. In the simplest method, the sinks are removed manually by flagging them in a preprocessing stage. This process is tedious and labor-intensive but may be the only option (as opposed to an automated technique) in certain cases. For example, in the case where a DTM is being created from a contour or spot map, a sink may be caused by an error in the assignment of an elevation value to a digitized contour line or spot data. In such a case, it is desirable to correct the elevation assignment. One way to detect these particular errors is to regenerate the contour lines after generating the DTM. Figure 8.4 shows the contour lines generated from a DTM with a sink. The sink is identified by the numerous concentric contour lines in the center of the map. Physically, according to this contour map, all land in the immediate vicinity of the sink drains (flowlines perpendicular to the contour lines) into the sink. The contour line map on the left identifies the location of the sink and allows the user to correct it. The line map on the right shows the same region with the correct spot data point.

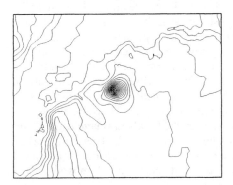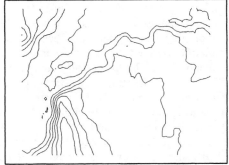

Figure 8.4 A contour line map with a sink (left) and with the sink removed (right).

Sinks and peaks caused by other types of errors associated with accuracy can be numerous in a DTM and consequently, a more automated approach is preferred. According to Raaflaub (2002), techniques can range from simply filling in sinks and smoothing out peaks to redirecting flowlines to neighboring cells. Several pit removal techniques require knowledge of the surface's aspect, and, by extension, its slope (Raaflaub, 2002).

The technique of filling in sinks and flattening out depressions is a very common approach to correcting DTMs for drainage purposes. A sink can be filled using the following procedure: (1) initiate a lake at the elevation of the sink cell, with a "shoreline" defined by the cell's perimeter; (2) find the lowest cell adjacent to the shoreline, raise the lake to that level, and expand the shoreline to include it; (3) if one of the neighbors is now lower than the lake, it is therefore the outlet, so terminate the process; and (4) if the lowest neighbor is part of another lake, merge the lakes and continue.

The problem, however, still arises as to how to direct flow once the sink is filled or peak is flattened. To deal with this issue, some algorithms allow water to flow between neighbors at the same elevation, determining the direction of flow by evaluating local slope (i.e., over a larger window). Alternatively, other algorithms deal with the problem by regarding the cell as a very small lake and simulating its overflow.

Note that while it is possible for sinks to be found in topographic surfaces and it may be necessary to retain them (such as in wetlands or "prairie potholes"), most sinks can be considered errors since fluvial erosion processes will not normally produce such features at the scale resolved by DTMs (Band, 1986 in Raaflaub, 2002).

8.2 CREATING HYDROLOGICALLY SOUND DTMS

Previous chapters in this text have provided methods for creating surfaces and DTMs, for example KRIGING. However, creating a surface of a terrain that drains according to physical laws is not always accomplished by a tool such as KRIGING. In order to use DTMs for hydrological modeling, the DTM must be *hydrologically sound*. This has been alluded to in Sections 8.1.1 through 8.1.2 with the notion of hydrologically connected regions and the drainage process. Figure 8.5 depicts a contour line map of a region in southern Alberta near Cochrane.

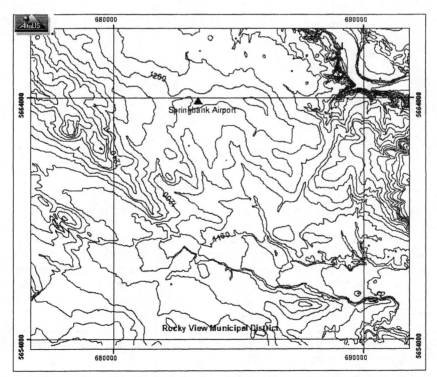

Figure 8.5 Contour line map produced from a portion of NTS Map Sheet 82O/1 (as seen in Figure 7.11) showing the Springbank Airport in Alberta. (Data courtesy of AltaLIS and all rights are reserved by the Government of Alberta, 2004.)

A DTM of 25m resolution was produced from this contour map using ordinary KRIGING and this is shown in Figure 8.6. A second DTM was created using the method of Hutchinson (1989) and this is shown in Figure 8.7.

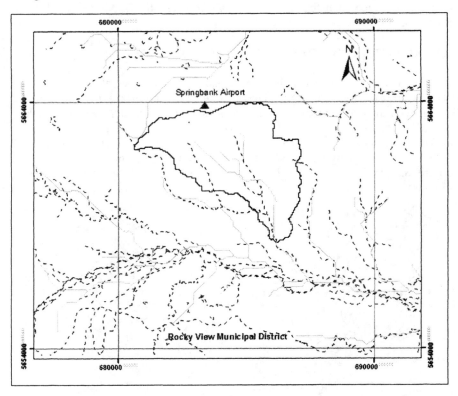

Figure 8.6 Stream network and watershed computed from a DTM developed using KRIGING.

Figure 8.6 shows the actual stream network (represented by a dashed line) superimposed with the software-determined stream network (represented by a light gray solid line). The solid dark polygon represents a watershed delineated using the DTM that drains in the southeast direction. The watershed boundary, which by definition bounds all water to flow toward the outlet, is shown to cross a river network that is physically unrealistic.

Figure 8.7, on the other hand, shows a drainage network created from a DTM that was created using a method based on Hutchinson (1989) and preserves realistic drainage patterns.

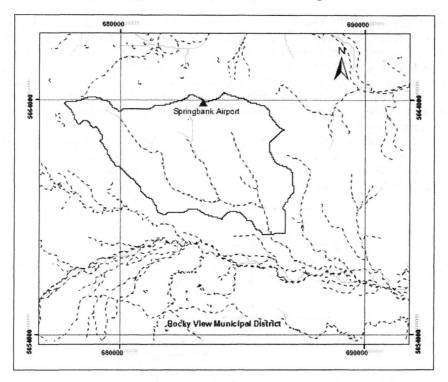

Figure 8.7 Stream network and watershed computed from a DTM developed using the method of Hutchinson (1989).

Notice that the predicted stream network (shown as the solid line) more closely matches the actual river network. Notice also now that the watershed is much larger and encompasses the Springbank Airport. This illustrates the importance of encompassing known physical processes in the creation of digital terrain models when they are being used for environmental applications such as delineating watersheds.

The method of Hutchinson (1989) has been implemented in many different types of software (ANUDEM and ARCGIS) and produces "sensible" drainage networks because it uses the stream network in a drainage enforcement algorithm that automatically removes spurious sinks. The algorithm is coupled with an iterative finite difference interpolation technique.

The enforcement algorithm as described in Hutchinson (1989) essentially works on the premise that any sink identified in the interpolation algorithm is surrounded by a drainage divide and that divide contains at least one *saddle point*. A saddle point is defined by a grid cell having at least two neighbors that are higher than the center cell. These are interwoven with other lower elevation neighbors. Figure 8.8 describes the notion of a saddle point. Saddles are linked to sinks through "ordered chain connections" that represent inferred flowlines. The linkage is made from sinks that lead to the lowest associated saddle point to the next lowest sink, which is found by looking in the steepest downhill direction away from each saddle point until the sink is found. This is illustrated in Figure 8.8 where the dashed line represents inferred drainage lines. The ordered chains impose a linear descent. The individual sinks are then cleared in order of increasing elevation.

Note that the type of river network shown on the right in Figure 8.2 is not permitted in the application of the method of Hutchinson (1989). The user must orient all the stream arcs in the downstream direction and split such loops using one arc and then remove the outside arcs.

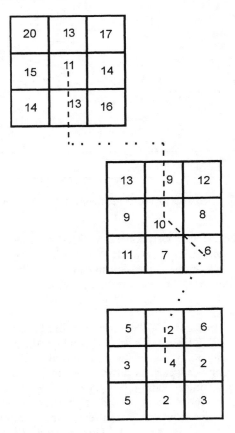

Figure 8.8 Drainage enforcement through sinks in saddles according to the method of Hutchinson (1989).

8.3 DETERMINING FLOW DIRECTION AND ACCUMULATION

Determining how and where water is flowing is naturally a first step in watershed delineation. The *flow direction grid* is a useful input for determining watershed boundaries and flow accumulation in each cell. Flow accumulation in a cell can lead to the delineation of a stream network.

Figure 8.9 depicts two flow direction grids where each cell is assigned a number indicating the direction of flow in either one of four (cardinal) directions (top figure) or one of eight directions (bottom figure). The direction is determined as the direction of steepest descent, which is determined by comparing the elevation between the center cell and its legitimate neighbors. The largest difference indicates the flow direction.

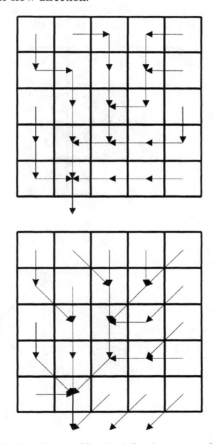

Figure 8.9 Flow direction indicated in one of four (top) directions or one of eight (bottom) directions.

Naturally, the direction of flow of water out of a cell is determined by the elevation of surrounding cells. A variety of algorithms exists for determining flow direction and at the most basic level, can take one of two cases: assuming only four possible directions of flow (up, down, left, and right); or assuming eight possible directions. In both cases water is assumed to flow from each center cell to the lowest of its neighbors (steepest descent, or highest gradient), and if no neighbor is lower, the cell must therefore be a sink.

In Figure 8.9, the shaded boxes indicated cells that have eight or more upslope cells draining into those cells. Notice that the location and number of gray cells for only four possible directions differs from that for eight directions. This has ramifications when determining flow accumulation (Section 8.4).

Figure 8.10 depicts a hypothetical DTM with elevations shown at the center of each cell. Figure 8.11 shows the direction number assigned to a center cell to indicate the direction of steepest descent out of the cell.

11	9	12	14
9	8	6	7
6	5	4	5
6	0	2	6

Figure 8.10 Digital elevation model.

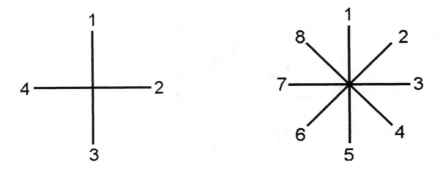

Figure 8.11 Direction numbers for a single flow in one of four (cardinal) directions (left) or one of eight directions (right).

Figure 8.12 shows the resulting flow direction grids from using the four or eight possibilities.

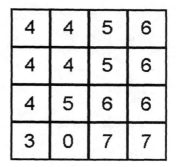

3	3	3	3
3	3	3	4
3	3	3	4
2	0	4	4

4	4	5	6
4	4	5	6
4	5	6	6
3	0	7	7

Figure 8.12 Flow direction based on four possible flow directions (left) or eight possible directions (right).

Regardless of whether direction is selected based on only four or eight possibilities, flow direction algorithms may be further subdivided as to whether they will allow flow in more than one of the possible directions.

8.3.1 Single-Flow Direction Algorithms

In both cases in Figure 8.11, flow was only allowed to exit a cell based on the steepest descent. These are thus termed *single-flow direction algorithms*. The single-flow direction based on eight moves is also sometimes referred to as D8 and was originally developed by O'Callaghan and Mark (1984). Hybrids such as the randomized single-flow direction method developed by Fairfield and Leymarie (1991) add a stochastic element to the flow direction. This algorithm produces a mean flow direction equal to the aspect, which tends to provide more realistic representations of flow (Gallant and Wilson, 2000). The problem with single-flow direction methods like this is that they tend to produce striated flow accumulation maps with "straight-line" artifacts (see Section 8.4). In response to the desire to produce more realistic-looking maps of flow proportion, *multiple-flow direction algorithms* were introduced that disperse the flow to more than one neighbor.

8.3.2 Multiple-Flow Direction Algorithms

Multiple-flow direction algorithms partition the flow to one or more downslope cells as opposed to just the direction of steepest descent as determined by single-flow direction algorithms. The differences can have a number of ramifications for how moisture is distributed in a watershed and this is further investigated in

Section 8.6. Figure 8.13 illustrates the difference between single- and multiple-flow direction algorithms.

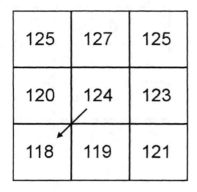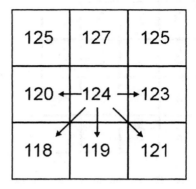

Figure 8.13 Difference between single-flow (left) and multiple-flow (right) direction algorithms.

The manner in which this flow is partitioned varies, but one example can be to partition flow based on the size of the slope between the center cell and its downslope neighbor, therefore having a slope weighting apportioning. Another method defined by Costa-Cabral and Burges (1994) is called DEMON. The DEMON method sends the flow generated at each source pixel through a stream tube until it encounters the DTM edge or a sink. According to Gallant and Wilson (2000), these stream tubes are not constrained by the structure of the grid (that is, to coincide with the edges of cells), and expand and contract as they traverse divergent and convergent regions of the DTM. The direction of flow (i.e., the stream tube) is based on aspect (as originally proposed by Lea, 1992) defined from triangular planar facets defined from the eight neighbors (Tarboton, 1997). The reader is directed to these references for further details.

Note that the method of assigning flow to more than one pixel assumes that dispersion is taking place. This is inconsistent with the definition of an area draining downslope to a single point (Tarboton, 1997). Flow direction methods are also more computationally intensive but tend to produce more realistic flow accumulation patterns.

8.4 FLOW ACCUMULATION AND STREAM NETWORKS

Flow accumulation in a grid cell results from upstream areas draining into downstream regions. This is a useful data layer to develop as it can indicate the location of stream networks under the assumption that stream networks are generated only at locations that reach a certain threshold of accumulation. To create a flow accumulation grid, the computer algorithm assigns the number of upslope cells (or total area = number of upslope cells × grid cell resolution) that flow into that cell. This involves tracking back from each starting cell up through the cells that flow into that cell, as indicated in the flow direction grid. All the cells that eventually flow into that cell are counted and added to the center cell which itself is counted as one. This is illustrated in Figure 8.14.

Both multiple- and single-flow direction algorithms may be used to determine flow accumulation. As mentioned earlier, single-flow direction methods tend to produce striated, straight-line artifacts in the DTM whereas multiple-flow direction algorithms do not have these features. This issue is more important in determining moisture or water accumulation through a DTM (see Section 8.6) than it is in the delineation of stream networks. See Section 8.6 on the topographic wetness index that further illuminates the striated nature of the flow accumulation.

Another important issue with regard to flow accumulation methods that rely on multiple flow directions is that the processing must take place in the order of descending elevation. That is, the processing must start with the most upslope cell and then proceed to the next downstream cell in the appropriate order. GIS systems that process data on a row-by-row basis regardless of the general direction of slope will only be able to conduct procedures based on single-flow directions. Multiple-flow direction analysis must generally be processed with external programs.

Determining stream networks helps to determine the location of the drainage point for the watershed. Since small quantities of water generally flow overland, not in channels, in natural systems, we may want to accumulate water as it flows downstream through the cells so that channels occur only when a threshold volume is reached. Thus a critical threshold value must be specified by the user in order to determine the location of a stream cell. This is illustrated in Figure 8.14.

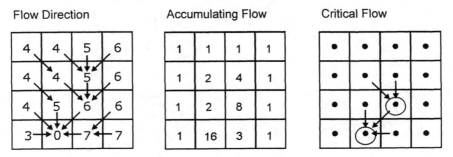

Figure 8.14 Determining the network using a critical threshold level of eight cells or more.

Both multiple- and single-flow direction algorithms tend to lead to similar stream networks but in the application of either, the important question is what critical threshold should be selected. Figure 8.15 shows a stream network produced for the region of Cochrane, Alberta (shown in Figure 7.11), when a critical threshold level is only 100 upslope cells. Figure 8.16 shows the stream network produced when the threshold is 1,000 cells.

Figure 8.15 Streamflow network for Cochrane, Alberta, region determined from using a critical threshold of 100 cells.

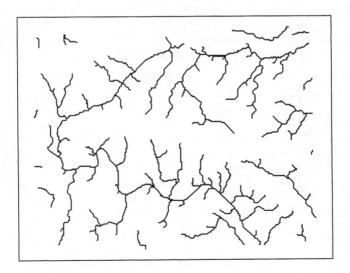

Figure 8.16 Streamflow network for Cochrane, Alberta, region determined from using a critical threshold of 1,000 cells.

The question then becomes, what is the appropriate threshold to use? If a threshold that is too large is selected, then the river network can become discontinuous and miss perennial (always flowing) streams of interest to the user. A threshold that is too small will produce an overly dense network of streams with streams that are not seen on any topographic map at any time of the year. The answer relies in the final application of the network. Tarboton et al. (1991) proposed that digital stream networks derived from elevation data be extracted at the "correct length scale" or drainage density for that basin. Their method uses physical scaling relationships of the kind developed by Horton (in Horton's law of basins; see Section 8.1.2) that dictate the minimum threshold that should be selected to extract the drainage network appropriate for that catchment.

If maps or aerial photos of the region are available, then stream networks may be compared to the original network observed on topographic maps, but if that is the case, then the original network should be used. However, occasionally this network does not exist and a digital network must be extracted from digital elevation data.

8.5 WATERSHED DELINEATION

A watershed is defined by its outlet; thus, before a watershed can be delineated, the user must specify the location of the outlet. This may be a hydrometric gauge that measures water level or the future location of a bridge where the designer is interested in knowing possible river water levels. In any case, to delineate a watershed boundary, a flow direction grid is required. Once the outlet (or outlets if subwatersheds are desired) has been specified, the flow direction grid then easily flags all cells eventually draining into the outlet. The flagged cells constitute the watershed. In Figure 8.17, the flow direction grid is shown in the upper left. Four outlets have been indicated with single dots in the outlet cells. The flow accumulation grid is shown in the figure in the top right and the resulting four subwatersheds (SB1 to SB4) are shown at the bottom of the illustration.

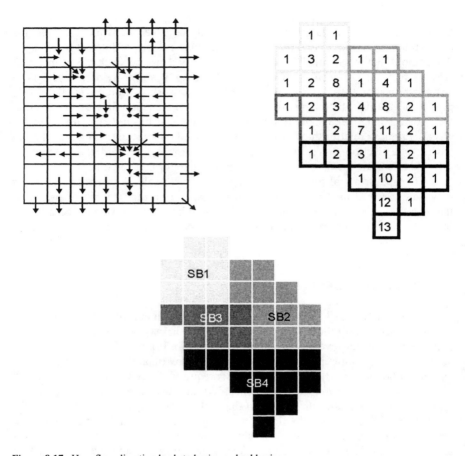

Figure 8.17 How flow direction leads to basins and subbasins.

8.6 THE TOPOGRAPHIC WETNESS INDEX

Models of hillslope hydrology endeavor to capture the numerous processes contributing to storm flow that occur at the hillslope scale. These processes are modeled by accurately accounting for the distribution of soil moisture in the catchment over time, which is heavily influenced by soil type and topography (Chorley, 1978). The *topographic wetness index* (Beven and Kirkby, 1979) is an index that has become extremely popular in hydrological and environmental modeling because it distributes soil moisture as a function of topography. Relatively simple in design, it is easily amenable to implementation with digital terrain models. The distribution of this index within a watershed essentially dictates the growth of the area contributing to surface runoff during a rain event.

The index is designed from three important assumptions that deal with topography and soil, and culminate into a model of potential surface saturation zones.

The topographic wetness index essentially provides a measure of the degree of saturation possible at a point in the terrain. The equation for the index is shown in (8.1):

$$\lambda_i = \frac{a_i}{\tan \beta_i} \tag{8.1}$$

If i refers to a single grid cell (with n being the total number of grid cells in the watershed), then a_i is the total upslope area that flows through that cell (that is the amount of flow accumulation to that cell) and β_i is the slope angle of that cell. Thus, the propensity of a point to saturate after a rain event (a high value of λ_i indicates a high propensity to saturate) is a function of flow accumulation and local slope. The lower the slope, the lower is the gradient to move water along and thus it accumulates more easily in the cell. Hence, the input of a DTM for the computation of this index is essential. Figure 8.18 shows a map of the topographic wetness index for the Ancaster Creek catchment. This catchment of roughly 8 km² in southern Ontario has a basin length of roughly 4 km. In the grayscale map, lighter values have higher values of the index and are more likely to saturate earlier after a rain event than darker regions. The few black polygons shown are flat regions that were eliminated from the computation.

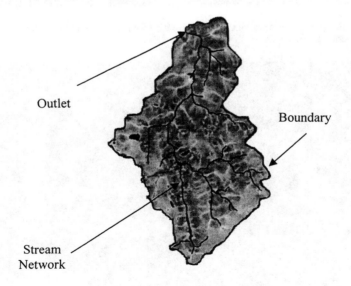

Figure 8.18 Topographic wetness index map of Ancaster Creek. (After Valeo, 1998.)

The index can indicate areas of possible saturation after a rain event and is generally computed from flow accumulation and slope. Thus, multiple- and single-flow direction algorithms can be used but are both these algorithms physically realistic? The effects on computing λ_i using single- or multiple-flow direction algorithms are shown in Figures 8.19 and 8.20. Close-ups of the same portion of the divide are shown in Figure 8.21.

Figure 8.19 Top portion of the Ancaster Creek catchment showing the wetness index created using a single-flow direction algorithm. (After Valeo, 1998.)

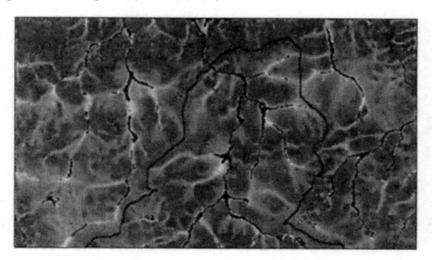

Figure 8.20 Top portion of the Ancaster Creek catchment showing the wetness index created using a multiple-flow direction algorithm. (After Valeo, 1998.)

 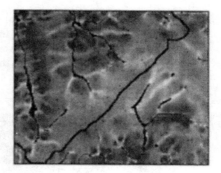

Figure 8.21 Single- (left) and multiple- (right) flow direction algorithm effects on topographic wetness index for the region near the divide showing the upslope region of the catchment. (After Valeo, 1998.)

In all figures, the whiter areas show high values of λ_i, which means that those areas in white are likely to have higher moisture levels than those that are darker. There has been some additional processing in the multiple-flow direction maps that involve removing flat areas which now appear black. In all the examples using the single-flow direction algorithm, the maps show a striated look while the multiple-flow direction algorithm produces a much smoother representation of the index. The striated look is caused by the fact that flow is channeled in single directions, whereas the multiple-flow direction algorithm disperses flow out among its downslope neighbors. In this case, the flow was dispersed based on a slope-weighted apportioning. If the index provides an idea of the soil moisture distribution in a watershed, then it is supposed that the multiple-flow direction algorithm provides a more physically realistic representation of this quantity.

8.7 TIN VERSUS GRID WATERSHED DELINEATION

The use of TINs for delineating watersheds received some attention when it was first introduced but waned in popularity in favor of a raster model for delineating watersheds. Recently however, TIN models have seen a resurgence in environmental modeling because of their ability to represent variability in terrain or other surfaces. For example, in flatter terrain, TINs represented the region with fewer and larger triangles as the terrain does not vary greatly with respect to elevation. Furthermore, raster grids which must maintain the same resolution throughout lose information in highly variable terrain that may affect drainage patterns.

Figure 8.22 shows an example of a TIN model for the Cochrane, Alberta region. This TIN was developed using less than 10,000 nodes. In a similar way in which DTMs are created with a resolution specified by the user, a user can specify the number of nodes to be used in the representation of the terrain for the entire region. Few nodes mean that less variability can be represented in the terrain as the triangles become larger because there are fewer that can be defined.

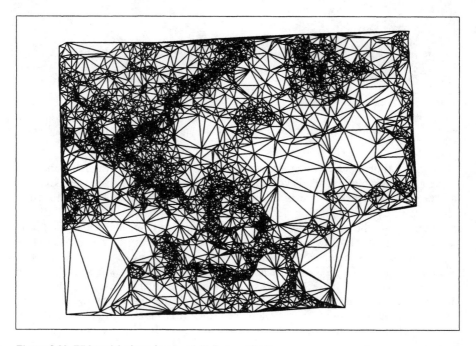

Figure 8.22 TIN model of terrain around Cochrane, Alberta.

As indicated earlier, water flow across a surface is primarily governed by the direction of steepest descent. Paths of steepest descent are identified in the TIN by starting at some selected point in a triangle and then computing the path of maximum gradient across that triangle, and then moving onto the next triangle in the same manner (Moore et al., 1991). The method of determining steepest descent over a TIN can provide the location of a stream network so long as the path of steepest descent coincides with the edges of triangles.

To delineate watersheds and subwatersheds, it is necessary to determine source areas contributing to a point in the system. The main problem is that some triangles will straddle the boundaries between source areas and thus lie in more than one source area. This is shown in Figures 8.23 and 8.24. Figure 8.23 shows a selected tributary and a grid of resolution 25m was used to delineate the watershed that this tributary drains. It is shown in gray in Figure 8.24. Figure 8.25 is a closeup view of the headwater areas of this subwatershed.

Small tributary
of watershed
shown in Figure
8.24

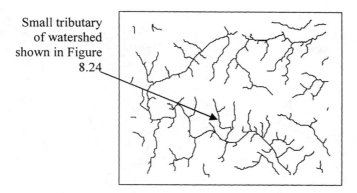

Figure 8.23 Stream network shown for Cochrane in Figure 8.16 and the selected tributary used to delineate watershed shown in Figure 8.24.

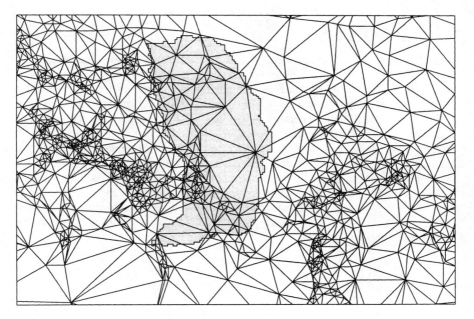

Figure 8.24 Tributary watershed superimposed on a closeup of the TIN surrounding the tributary.

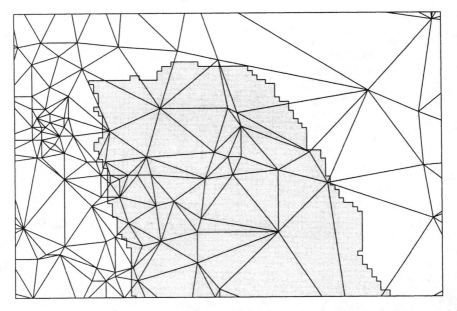

Figure 8.25 Close up of the headwater area (area near the divide) for the tributary's watershed.

Notice that the watershed as defined by the raster grid only constitutes subareas of the triangles near the divide. This is an indication that the number of grid nodes is insufficient to capture the drainage characteristics of this region. The solution is therefore to increase the number of nodes. Figure 8.26 shows a portion of the TIN near the headwater area and a probable path of steepest descent toward the outlet. The triangles near the divide are currently too coarse to capture the location of the divide properly and it is unknown where the path of steepest descent occurs. The solution, which is shown in the figure on the right-hand side of Figure 8.26, is to subdivide each triangle so that each smaller triangle contributes directly to flow on only one side of the divide. This is shown in the bottom right section of the figure.

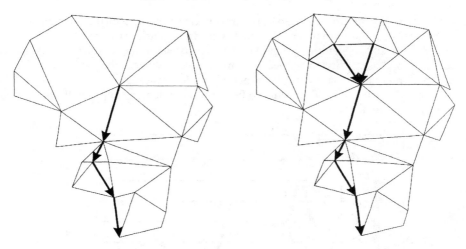

Figure 8.26 Small portion of TIN near headwater area and probable path of steepest descent.

Once the contributing areas are properly calculated, they can be used to delineate watersheds. Using TINs for watershed or stream network delineation does have problems, including the possibility of divergent streams (a stream actually split in two while flowing downstream), flat channels, or flat channels and sinks.

8.8 ISSUES OF SCALE AND ERROR

An important issue that has only been alluded to in Section 8.4 is the issue of resolution and scale when delineating watersheds and drainage networks. What are the appropriate grid resolutions when delineating watersheds and what are the critical thresholds to select in determining a drainage network when dealing with an area at the 1:50,000 scale, for example? The issue of resolution is an important one as it has an impact on numerous modeling parameters that are derived from terrain. A watershed boundary is easily delineated from a DTM but the parameter that is most often used in hydrological modeling is the area of the watershed, not the physical location of the boundary. Figure 8.27 shows the relationship between watershed area and grid resolution for a variety of resolutions. The finest scale possible for delineating a watershed given the scale of the input data used to create the DTM was a grid resolution of 10m. The DTM was created to delineate the Ancaster Creek catchment shown in Figure 8.18 and was created using the method of Hutchinson (1989). DTMs were then created using the same method but with higher resolutions and the same watershed area was delineated.

As the figure shows, the percent difference from the base case (10m) can fluctuate by as much as 7% with regard to cell size chosen. This has serious implications for runoff generation programs because the size of the watershed is directly related to the amount of flow that is generated during a storm.

Notice also a decreasing trend in area. This is due to the fact that as a larger cell size is chosen, the terrain becomes much smoother near the divide and this results in a greater area being determined to reside outside the divide as opposed to within the watershed. This can also be seen in the demonstration with the TIN shown in Figure 8.25.

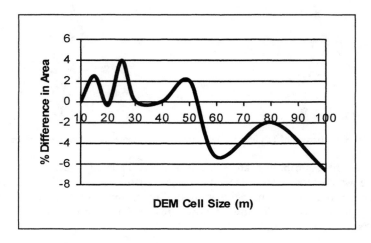

Figure 8.27 Effect of cell size on basin area. (After Valeo, 1998.)

Lee and Chu (1996) tested the effects of errors on DTMs and their results showed that errors caused a reduction in the number of drainage cells (a cell residing on a stream network). That is, the greater the number of errors that the DTM had, the fewer the number of cells that equaled or exceeded the accumulation threshold that rendered it a stream cell. Raaflaub (2002) indicated that DTM errors had significant implications for the computation of slope and aspect and thus could lead to significant implications for models relying on these parameters.

8.9 APPLICATIONS TO ENVIRONMENTAL MODELING

8.9.1 Surface Water Modeling

Hydrologic response units or HRUs (Wolfgang, 1995) are an excellent example of how subunits are used to determine flow characteristics. The hydrological response unit is a classic example of overlay analysis for land phase parameter estimation in which one of the layers of data being overlaid and fused with other data is a DTM. An HRU is a modeling subunit in which the runoff response to precipitation is homogenous within the subunit. Hydrological response, or runoff response, is a function of several factors including vegetation, soil type, and topography. Figure 8.28 shows the overlay process in which an HRU is uniquely identified as the unique combination of vegetation cover (with greater densities of vegetation and canopy generally creating less runoff), soil type (with higher infiltration soils creating less runoff), and topography (with steeper slopes creating faster runoff and dependencies on aspect).

The issue here is how many classes of topography one must create in order to merge a continuous variable such as topography with thematic data like landcover or soil type. Note that the final scale of the resulting HRU map is a function of the layer with the smallest scale (the coarsest resolution). The topography can be grouped into varying classes depending on the variability in the landmass or as a function of the software processing capabilities as the higher the number of classes, the greater the number of possible HRUs. Also, topography is not only used to generate classes of elevation, but it would also possibly lead to classes of slope and classes of aspect. In this case, the elevation itself is often used to assist in determining changes in temperature or pressure. When dissecting areas of topography, the primary issue is to examine the inherent variability in the dataset and determine how slopes evolve. The best way to do this is to initially start with many fine resolution classes and then statistically determine the amount of area that resides in each class. Classes should absorb adjacent classes that contain very little area in order to reduce the number of HRUs. Overlay exercises in which DTMs are a component in the overlay processes are used again in later sections.

8.9.2 Applications in Hydraulics

Once the quantity of flow is determined at a point, it is often necessary to *route* or move this flow downstream, and therefore predict its spatial and temporal qualities because surface roughness and terrain gradients either accelerate or slow down flow. One example of a routine method uses the kinematic wave routing equations. An example of a typical equation used in hydraulics is of the form:

$$\frac{\partial d}{dt} + \frac{\partial}{\partial x}\left(\frac{\sqrt{s}}{n} d^{5/3} \cos\alpha\right) + \frac{\partial}{\partial y}\left(\frac{\sqrt{s}}{n} d^{5/3} \cos\beta\right) = R - I - E \qquad (8.2)$$

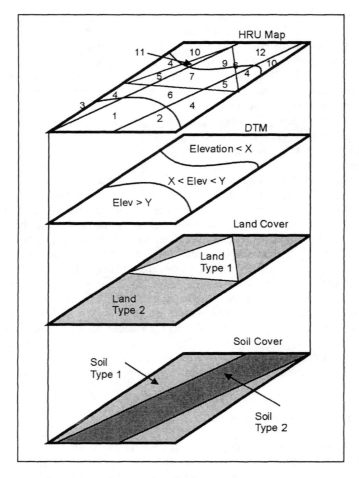

Figure 8.28 HRUs derived from soil type, land cover type, and topography.

Equation (8.2) is a typical physically based hydrological model's 2-D kinematic wave equation for routing overland flow. This equation essentially describes the depth of flow as a function of time and space over a piece of terrain given certain factors. The depth d is the model output, s is the slope of the terrain surface, n is a surface roughness factor known as *Manning's n*, α and β are direction vectors of overland flow, I is the infiltration rate of rainfall into the soil, R is the rainfall intensity hitting the land surface, and E is a loss of rainfall in the form of potential evaporation.

The equations in this model are solved using finite difference techniques that essentially solve partial derivatives by substituting them with finite difference representations on a grid of discrete points. What is essential here is the direction of flow, and the slope, which are both derived from DTMs. If a regular raster grid is used to input elevation (and thus derive direction and slope), the finite difference method of numerical analysis can be employed to solve the equation. Finite elements, which are suitable in cases where there is irregular geometry, unusual boundary conditions, or a nonhomogenous composition, are better handled when the DTM is in the form of a TIN.

In addition to this physically based equation for "moving" water across the land surface, the field of water resources engineering has numerous other equations for determining the quantity and timing of flow across a terrain surface. *Unit hydrograph* and *time-area* methods are tools that engineers use just for this purpose and they include a number of parameters derived as functions of elevation such as *time to peak* (t_p).

The time to peak is the time between the centroid of the rainfall event and the peak discharge resulting from that rain event. There are numerous equations for deriving t_p but as they are functions of the length of the basin and the slope, one suggested form (Watt et al., 1991) takes the form of

$$t_p = 0.000326\left(\frac{L}{\sqrt{S}}\right)^{0.79} \tag{8.3}$$

where t_p is the time to peak in hours, L is the length of the basin in meters and S is the slope of the basin in meters/meters. Digital terrain models have become indispensable in determining these lengths and slopes for numerous areas at once. The peak discharge and depth of flows are then determined as functions of t_p.

8.9.3 Erosion

Erosion is one of the most prevalent mechanisms for creating land formations. But erosion is also considered one of the most important and challenging problems in natural resource management in both developing and developed nations (Ward and Elliot, 1995). It destroys valuable farmland and fills streams and reservoirs. Eroded sediment can carry pesticides and nutrients, particularly phosphates that can lower water quality. Soil erosion also reduces the productivity of some soils by removing organic matter and degrading soil structure. Erosion can occur by both water and wind mechanisms but erosion by water is one of the most prevalent problems leading to nonpoint source pollution. In fact, erosion by water is considered one of the most widespread types of soil degradation and occurs to some extent in all provinces in Canada (Wall et al., 1997).

There are two general forms of erosion: natural and anthropogenic. Natural erosion involves the removal and formation of soil and occurs at geological timescales to produce canyons, streams, and valleys. Anthropogenically induced erosion tends to be an accelerated form of erosion and leads to loss in soil productivity. Activities leading to such soil loss can be the result of poor farming and tillage practices and clear cutting.

All types of erosion are affected by climate, soil type, vegetation, and topography. Climate provides elements of rainfall intensities and amounts. Soil texture is the most dominant property determining the erosion potential over a region. Smaller particles such as those in clayey soils are more difficult to detach than larger, sandier particles (but the smaller particles are more easily transported). Vegetation reduces erosion by protecting the soil from rain and reducing surface runoff velocity. The root system also holds soil in place and increases transpiration rates.

The topographic characteristics that influence erosion are slope, steepness, length, and shape (Ward and Elliot, 1995). Steeper, longer, and convex slopes are more erosive than milder sloping, shorter, and concave slopes.

Estimating soil erosion and loss is an essential part of resource management and one of the first and still most widely used methods for estimating erosion is the universal soil loss equation or USLE (Wischmeier and Smith, 1978). The method has undergone several modifications to include the revised universal soil loss equation (RUSLE) and the modified universal soil loss equation (MUSLE). RUSLE is very similar to USLE but provides improved estimates on the topographic factor. It is now a widely accepted method of estimating sediment loss on average over a long period of time for a region. It is useful for determining the adequacy of conservation measures in resource planning and for predicting nonpoint sediment losses in pollution control programs. MUSLE actually determines sediment yield for a specific rain event.

USLE and RUSLE take the following form:

$$A = R \times K \times LS \times C \times P \qquad\qquad (8.4)$$

A is the average annual soil loss in tons/acre. R is a rainfall erosivity index and varies with the amount of rainfall and the individual storm precipitation patterns. It is derived for various regions and is a function of the overall energy of a storm. The soil-erodibility factor K varies to account for seasonal variation in soil erodibility with higher erodibility in the spring and/or after tillage. The topographic factor LS provides greater erosion rates on longer and/or steeper slopes and lesser erosion on flatter or shorter slopes. The cover management factor C accounts for effects of vegetative cover, crop sequence, productivity level, length of growing season, tillage practices, and residue management. For disturbed bare soil, a value of 1.0 or greater should be used. Erosion from permanent pasture, rangeland, and forest is generally much lower than from agricultural lands. Human and/or livestock activities that disturb the vegetation, such as roads, grazing, or timber harvest are generally the source of most of the eroded sediment from rangeland and forests. Finally, vegetation cover or farming practices have major effects on soil erosion rates. Erosion control practices include contouring, strip cropping, and terracing, and these are embodied in the P factor.

Kienzle (1996) demonstrated that DTM grid mesh size greatly impacted slope values such that increasing grid size meant decreasing slopes. This translated into a difference of annual soil loss potential of over 150% when using a grid size of 400m instead of 50m. The impact on the RUSLE was to decrease the amount of soil loss when increasing the grid cell size (as this has the net effect of smoothing slopes and misrepresenting DTM features).

Quinn and Anthony (1999) noted that modeling the washoff of diffuse surface pollutant sources and sediment requires an accurate representation of processes that are highly localized and controlled by small-scale natural and anthropogenic-made features. Hence, DTM scale is vital to the accurate representation of the flow pathways mobilizing and transporting these pollutants.

8.9.4 Fire Risk Modeling

While some of the previous applications show parameters that can be derived solely from DTMs, what we observe in the environment is a complicated combination of the interaction of various phenomena and how they are influenced by terrain (and in return, how they influence the formation of terrain). Geographic information systems have become valuable resources for just these kinds of operations as they allow for merging of multiple datasets. Very often in understanding conditions of the environment over terrain, the terrain itself is a necessary input that needs to be combined with other sources of information. An excellent example of this is the HRU discussion shown in Section 8.9.1. Here various kinds of data layers are combined in an *overlay* approach where the user must merge layers according to a mathematical expression that results in a single number. This number can be a risk factor, or a simply an index that is classified into pragmatic classes.

An additional example of this is similar to the HRU approach and can be shown in highlighting areas of fire risk. In many cases, fire risk is defined by the direction of prevailing hot winds, by the slopes that trend up from that direction, and from the vegetation biomass on those slopes. Slopes facing the prevailing wind and that have high vegetation biomass are at high risk. Also, steeper slopes are at higher risk, as the flame front can travel quickly up a steeper slope. In this case, we have three factors to take into account: slope (from the DTM), aspect (from the DTM), and vegetation mass (from satellite imagery, e.g., Landsat TM). In this case, we would like to express fire risk as a number from 0 to 1, with 1 being the maximum fire risk. In the following example, considering that the prevailing hot wind direction is S-N and looking at each of the three factors, the formula would be made up of the following components: slope, aspect, and a vegetation index.

When considering slope, a slope of $90°$ would be the most dangerous, and a slope of zero would be the lowest. To reduce this to a portion of the formula, we would simply divide the slope (in degrees) by 90 in order to normalize each index from 0 to 1.

$$\frac{\text{Slope}}{90} \tag{8.5}$$

With regard to aspect, slopes facing (heading) south have the highest fire risk, while slopes facing north have the lowest. We express this as a number from 1 (high) to 0 (low) fire risk as follows:

$$\left(1 - \left|\frac{180 - \text{Aspect}}{180}\right|\right) \tag{8.6}$$

Finally, with regard to vegetation, a normalized difference vegetation index (NDVI) value is a suitable index to provide information on the amount of biomass present in the area. It is already normalized such that it ranges from -1 to 1, with 1 being high vegetation biomass and -1 being the lowest. We reduce this to a number from 1 (high risk) to 0 (low risk):

$$\frac{\text{NDVI} + 1}{2} \tag{8.7}$$

Putting all this together, we get a generic formula, adding these three factors together and dividing by three:

$$\text{Risk} = \frac{\dfrac{\text{Slope}}{90} + \left(1 - \left|\dfrac{180 - \text{Aspect}}{180}\right|\right) + \dfrac{\text{NDVI} + 1}{2}}{3} \tag{8.8}$$

This will result in a fire risk layer being output, with 1 as high fire risk, and 0 as low fire risk. Figure 8.29 shows the original DTM and the computed fire risk map on the left.

We could drape the above fire risk layer over the original DTM as a *height layer*, and view the processed result in 3-D. We could also choose to change the layer to a classification layer, and apply a threshold to the final transform (perhaps only selecting regions with a greater than 80% fire risk). This is shown in Figure 8.30.

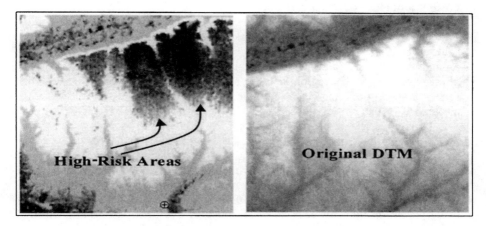

Figure 8.29 Original maps and highlighted areas.

The fire risk classification layer, which is computed from the original data, could be converted into vector polygons for input into a GIS system. Before doing so, we might choose to run a median filter to remove small regions over the image (this filter would be inserted into the layer after the above formula). A median ranking filter will reject small regions of fire risk, ensuring that we only end up with large fire risk polygons for the GIS.

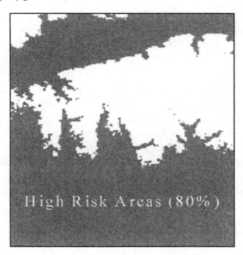

Figure 8.30 Identified high-risk areas of greater than 80% shown in white.

8.9.5 Avalanche Risk Management

Avalanches are caused by the combination of several environmental conditions. Factors contributing to the occurrence of avalanches include the physical properties of the area and meteorological conditions (Colpitts et al., 2004). The physical properties of a region that lead to avalanches include topological factors such as slope, aspect, and curvature. Since all of these quantities are derived from terrain, DTMs can play a vital role in avalanche early warning systems.

8.9.5.1 Developing Avalanche Zone Terrain Parameters Using DTMs

Of all the topographical factors influencing the occurrence of an avalanche, slope by far is considered one of the most important influences on the risk of an avalanche occurring. The slope angle of the terrain determines the amount of force being placed on the snowpack and how much of this force is in the downslope direction (Colpitts et al., 2004). As the slope angle increases, the force pushing on the snowpack increases. Because snow accumulation is smaller on slopes with higher inclination angles, these larger slopes will induce a greater number of smaller avalanches as opposed to fewer larger avalanche events. The greatest number of avalanches occurs between slope angles of 30° and 60°; any angles less than 30° are not amenable to avalanche initiation, and angles greater than 60° tend to create very small avalanches that make regular deposits of small amounts of snow. Furthermore, data collected in Canada and Switzerland seem to indicate that the majority of avalanche events occurred on slopes between 35° and 45° (Colpitts et al., 2004). It is a common misconception that avalanches can only occur at greater slope angle. There are certain circumstances in which avalanches can start on any degree of slope (with the exception of slopes that are forested).

The aspect of a slope is also a critical factor in determining the stability of the snowpack. For certain exposures, the slope may be exposed to increased levels of wind that contribute additional windswept snow to the snowpack. This wind loading often causes the snowpack to become unstable and more susceptible to avalanches (Colpitts et al., 2004). The aspect of the slope also determines the amount and duration the snowpack is exposed to solar radiation. For certain aspects with greater solar exposure, the snowpack will experience a greater fluctuation in temperature and this will impact the characteristics of the snowpack.

The curvature of a slope also plays an important part in determining the point along the slope where the avalanche begins (starting zone). Starting zones can exist along a variety of curvature include a convex slope, a planar slope, a gully, a ridge, or a rocky base. In many cases, the starting zone is either a planar or convex slope (Jamieson and Geldsetzer, 2003).

Vegetation and ground cover also influence the size of the avalanche that is possible. Trees in forested areas along the slope can act as anchors that hinder the movement of snow along the slope. Rocky areas with no vegetation provide little resistance to snow movement. Most avalanche events are initiated above the treeline but it is still possible for smaller avalanches to occur below the treeline.

Given that slope, aspect, and curvature are all factors influencing avalanche risk zones, DTMs are certainly vital in developing risk management plans during avalanche season. Colpitts et al. (2004) developed an avalanche risk index based on factors influencing the occurrence of avalanches in a similar manner to the fire risk index developed in Section 8.9.4. They developed the risk index for the Kananaskis region in southwestern Alberta (see Figure 8.31). The Kananaskis region is on the leeward side of the Rocky Mountains and contains a wide range of slopes, experiences significant snowfalls, and experiences a large number of hazardous avalanches every year. The DTM was created at a resolution of 25m using ARCGIS. Colpitts et al. (2004) developed an index of avalanche risk for this region by starting with slope, aspect, and vegetation cover. Figures 8.32 and 8.33 show the slope and aspect maps, respectively, for the region. Weights were assigned to each slope class and each aspect class in order to provide an indication of which types of slopes and which types of aspect would be greater contributors to avalanches than others.

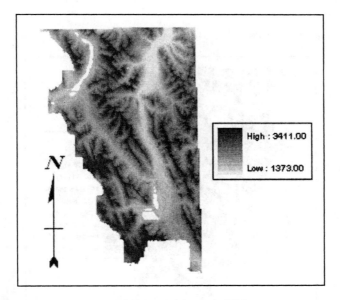

Figure 8.31 DTM used in avalanche risk index (Colpitts et al., 2004).

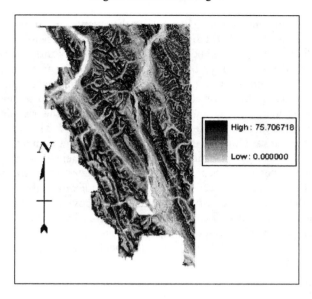

Figure 8.32 Slope map (Colpitts et al., 2004).

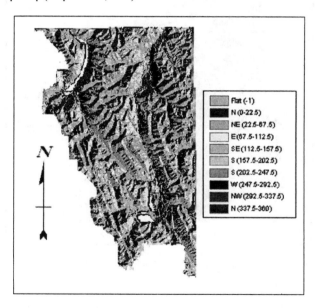

Figure 8.33 Aspect map (Colpitts et al., 2004).

8.9.5.2 Meteorological Parameterization Using DTMs

Both precipitation and temperature are critical meteorological factors affecting avalanche occurrence. The type of precipitation occurring (snow, sleet, rain and so forth), the size and shape of the precipitate, and the rate and duration of the precipitation are all important considerations. Large amounts of dense snow can accumulate on the snowpack causing sudden increases in snow load. Wind may further exacerbate the problem by creating snow accumulations in several smaller areas, thereby increasing the instability over a smaller area. Precipitation varies with elevation and in order to distribute the precipitation data from only a few meteorological stations spread out through the region, a method was required that utilized elevation from the DTM. Colpitts et al. (2004) used an inverse distance weighting scheme of the following form:

$$r_x = \frac{\sum_{i=1}^{n} (D_i / \Delta E_i)^b r_i}{\sum_{i=1}^{n} (D_i / \Delta E_i)^b} \tag{8.9}$$

where n is the number of meteorological stations; D_i is the distance from station i to point x; ΔE_i is the elevation difference between station i and point x; b is the order of inverse distance weighting (2 in this application); r_x is the interpolated precipitation for point x; and r_i is the precipitation at station i. In this way, precipitation can be interpolated between elevations and stations using the DTM for every cell r_i. The precipitation was then accumulated over a period of time beginning from the start of the snowfall season to the end of the snowmelt period.

Temperature is important in determining the stability of the snowpack as air temperature impacts snow conditions. Above freezing temperatures alter the condition of a snowpack by redistributing moisture within the snowpack heterogeneously. The snow ripens, becomes denser, and causes the snowpack to settle (Colpitts et al., 2004). Temperatures falling below freezing will freeze any moisture, making the snowpack vulnerable to fractures. In order to distribute temperature as a function of elevation to every cell in the DTM, Colpitts et al. (2004) used the following expression:

$$T_p = T_1 + \frac{Z_p - Z_1}{Z_2 - Z_1}(T_2 - T_1) \tag{8.10}$$

where T_p is the interpolated temperature at point p; Z_p is the elevation at point p; and $T_{1,2}$, $Z_{1,2}$ are the temperatures and elevations at the reference stations, respectively. A cumulative sum of the number of days that the maximum temperature rose above zero was calculated over the same period as the precipitation cumulative sum.

Note that wind plays a key role in the loading of additional snow on leeward slopes. The amount of wind the slope is exposed to is highly dependent on the location and elevation of area. But due to the highly variable nature of wind and errors in measurement, wind was excluded from the model. Also excluded was a characterization of snowpack profile and stability. When trying to determine high risk areas of avalanches, the most useful information comes from an analysis of the existing snowpack (Colpitts et al., 2004). Field tests assign a certain rating to a snowpack's stability. However, these are localized tests that are not necessarily representative of the surrounding snowpack. Hence, snowpack stability as a function of conventional stability ratings was also excluded from the model.

In the same way that the fire risk map was developed from (8.5)-(8.8), avalanche risk index was developed in a very similar way:

$$RF_i = \left(\sum_i P\right)\left(\sum_i T - Flux\right) S_i A_i V_i \tag{8.11}$$

where RF_i is the total risk factor for a pixel i; the sum over P is the cumulated precipitation over the snowfall period; the sum over T-$Flux$ is the cumulated sum of flux days over the snowfall period; S_i is the slope weight for pixel i; A_i is the aspect weight for pixel i; and V_i is the vegetation weight for pixel i. An avalanche risk map was created for March 22, 2001, and compared with two observations of avalanches on that day. This is shown in Figure 8.34.

The two observations for March 22 both occur in moderate-risk areas. Notice that there are very few high-risk areas for this region (as determined by the model). The comparison shows that the model is behaving reasonably; however, wind and snow stability were not incorporated in the model.

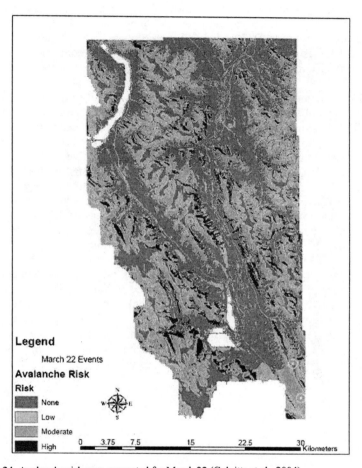

Figure 8.34 Avalanche risk map computed for March 22 (Colpitts et al., 2004).

8.9.6 Geomorphological Applications

Geomorphology is the study of the Earth's landforms and the processes that create them (Huggett, 2003). The erosion discussed in Section 8.9.3 is a process leading to landform creation. Ice sheet dynamics is another significant process landscape geomorphology. A persistent problem in studying landscapes formed by Pleistocene ice sheets is the inability to separate the landforms from the pervasive cultural and biological overprint. This is further complicated by the wide range of scales over which the landforms exist. In the past, most mapping and interpretation of glacial landforms has been conducted using aerial photographs and more recently satellite imagery. However, due to their immense size, many landforms are not detected using these data. With an increase in the availability of low- and high-resolution DTMs, it has become possible to exclude the cultural "noise" and thereby properly identify and interpret the geomorphic context of landforms at different scales (Sjogren, personal communication with the author, 2004).

Reconstruction of glacial environments relies heavily on sedimentary data that is only useful when it can be put into a landform and regional context. The use of DTMs for visualization and geomorphometric characterization and classification is fundamental for establishing this context. Various derivatives of elevation, particularly hillshade models, are the most effective for visualizing the cross-cutting relationships of landforms on local and regional scales (Sjogren et al., 2002). The glaciated landscape illustrated in Figure 8.35 clearly shows the complexity of glaciated landscapes and the power of simple hillshade models to place these landforms into a relative chronological framework (Sjogren, personal communication with the author, 2004).   The hillshade representation of a region in Alberta was developed from a 3 arc second SRTM projected to an Albers equal area at 100m postings.

Wildfires can also have significant influences on landscape formation and ecology (Martin, personal communication with the author, 2004). DTMs are used in conjunction with sophisticated mathematical models to predict shallow landslide formation after wildfires over thousands of years. Figure 8.36 depicts modeled erosion and deposition after a 10,000 year period for a region in coastal British Columbia, Canada. The 3-D rendering provides visual information on the types of slopes experiencing erosion versus deposition.

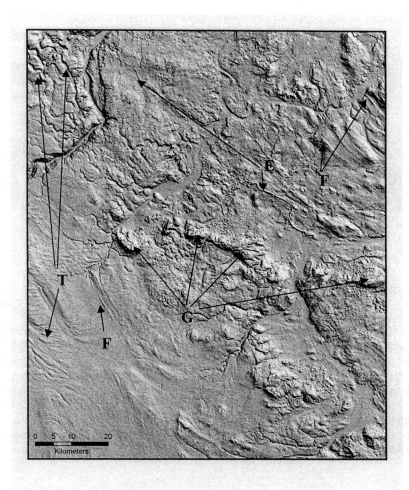

Figure 8.35 Hillshade indicating a number of landforms including fluting (F), eskew (E), glaciotectonic ridge (G), and subglacial channels (T). (Photo courtesy of D. Sjogren, Dept. of Geography, University of Calgary.)

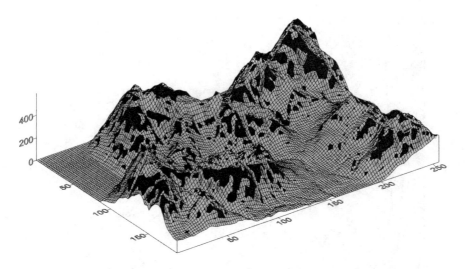

Figure 8.36 A 3-D rendering of modeled deposition (light areas) and erosion (dark areas) for a coastal British Columbian hillside. (Photo courtesy of Y. Martin, Dept. of Geography, University of Calgary.)

This chapter was intended to provide some idea of the range of possibilities and applications for DTMs in the environmental modeling. These are just the tip of the iceberg as DTMs are used in a whole host of other areas such as glaciology, soil vegetation atmosphere schemes for Earth system modeling, and oil and gas exploration. DTMs are used to develop simple empirical indices such as those shown in the fire and avalanche risk modeling, as well as theoretically derived parameters that are used in hydrologic and hydraulic modeling. The user must be aware of the end application of the DTM in the environmental modeling exercise in order to determine the impacts of errors and the necessary scale.

References

Band, L.E., "Topographic partition of watersheds with digital elevation models," *Water Resources Research*, 22(1):15–24, 1986.

Beven, K.J., and Kirkby, M.J., "A physically-based variable contributing area model of basin hydrology," *Hydrological Sciences Journal,* 24(1):43-69, 1979.

Chorely, R.J., "The hillslope hydrological cycle," in *Hillslope Hydrology*, Kirkby, M.J. (ed.), New York: John Wiley & Sons, 1-42, 1978.

Colpitts, C., Colpitts, J., Dixon, R., and Wong, S.B., "Avalanche risk modelling," *Geomatics Engineering, ENGO 500 Project*, University of Calgary, Calgary, AB: 49, 2004.

Costa-Cabral, M., and Burges, S.J., "Digital Elevation Model Networks (DEMON): A Model of Flow Over Hillslopes for Computation of Contributing and Dispersal Areas," *Water Resources Research,* 30 (6): 1681-1692, 1994.

Fairfield, J., and Leymarie, P., "Drainage networks from grid digital elevation models," *Water Resources Research,* 27: 709-717, 1991.

Gallant, J.C., and Wilson, J.P., "Primary topographic attributes," *Terrain Analysis, Principles and Applications,* Wilson, J.P., and Gallant, J.C., (eds.), New York: John Wiley & Sons, 2000.

Horton, R.E., "Drainage basin characteristics," *Transactions American Geophysical Union,* 13: 350-361, 1932.

Huggett, R.J., "Fundamentals of Geomorphology," *Routledge Fundamentals of Physical Geography,* London and New York: Taylor and Francis Group, 2003.

Hutchinson, M.F., "A new procedure for gridding elevation and stream line data with automatic removal of spurious pits," *Journal of Hydrology,* 106: 211-232, 1989.

Kienzle, S.W., "Using DTMs and GIS to define input variables for hydrological and geomorphological extraction," *HydroGIS 96: Application of Geographic Information Systems in Hydrology and Water Resources Management, Proceedings of the Vienna Conference,* IAHS Publ. No. 253: 183- 190, April 1996.

Jamieson, B., and Geldsetzer, T., *Avalanche Accidents in Canada, Volume 4, 1984–1996,* 2003.

Lea, N. L., "An aspect driven kinematic routing algorithm," in *Overland Flow: Hydraulics and Erosion Mechanics,* Parsons, A.J., and Abrahams, A.D., (eds.), New York: Chapman & Hall, 1992.

Lee, J., and Chu, C.J., "Spatial structures of digital terrain models and hydrological feature extraction," *HydroGIS 96: Application of Geographic Information Systems in Hydrology and Water Resources Management, Proceedings of the Vienna Conference,* IAHS Publ. No. 253: 201-206, April 1996.

Moore, I.D., Grayson, R.B., and Ladson, A.R., "Digital terrain modeling: a review of hydrological and geomorphological, and ecological applications," *Hydrological Processes* 5: 3-30, 1991.

O'Callaghan, J.F., and Mark, D.M., "The extraction of drainage networks from digital elevation data," *Computer Vision Graphics Image Processes,* 28: 328-344, 1984.

Quinn, P., and Anthony, S., "The use of high resolution digital terrain models to represent wash-off in natural and man influenced environments," *Impact of land-use change on nutrient loads from diffuse sources, Proceedings of the IUGG 99 Symposium HS3,* Birmingham, IAHS Publ. No. 257: 255-263, July 1999.

Raaflaub, L.D., "The Effect of Error in Gridded Digital Elevation Models on Topographic Analysis and on the Distributed Hydrological Model TOPMODEL," M.Sc. Thesis, Department Of Geomatics Engineering, University Of Calgary, Calgary, Alberta, UCGE Report No. 20163: 113, 2002.

Sjogren, D., Fisher, T.G., Taylor, L.D., Jol, M.J., and Munro-Stasiuk, M.J., "Incipient tunnel channels," *Quaternary International,* 90: 41-56, 2002.

Strahler, A.N., "Quantitative analysis of watershed geomorphology," *Transactions American Geophysical Union,* 38: 913-920, 1957.

Tarboton, D.G., Bras, R.L., and Rodriguez-Iturbe, I., "On the extraction of channel networks from digital elevation data," *Hydrological Processes,* 5: 81-100, 1991.

Tarboton, D.G., "A new method for the determination of flow directions and upslope areas in grid digital elevation models," *Water Resources Research,* 33(2): 309-319, 1997.

Valeo, C., "Variable Source Area Modeling in Urban Areas," Ph.D. Dissertation, McMaster University, Hamilton, ON, 1998.

Wall, G.J., Coote, D.R., Pringle, E.A., and Shelton, I.J., (eds.), *RUSLEFAC Revised Universal Soil Loss Equation for Application in Canada: A Handbook for Estimating Soil Loss from Water Erosion in Canada*, Research Branch Agriculture and Agri-Food Canada, Ottawa, Ontario, ECORC Contribution Number 02-92: 131, 1997.

Ward, A.D., and Elliot, W.J., (eds.), *Environmental Hydrology*, Boca Raton, FL: CRC Press, 1995.

Watt, W.E., Lathem, K.W., Neill, C.R., Richards, T.L., and Rousselle, J., "Hydrology of Floods in Canada", Ottawa: National Research Council, 1991.

Wischmeier, W.H., and Smith, D.D., *Predicting Rainfall Erosion Losses: A Guide to Conservation Planning*, USDA Handbook, Washington, D.C.: 537, 1978.

Wolfgang-Albert Flugel, "Hydrological response units (HRUs) to preserve basin heterogeneity in hydrological modeling using PRMS/MMS — Case study in Brol basin, Germany," *Modeling and Management of Sustainable Basin-Scale Water Resource System* (Proceeding of Boulder Symposium), IAHS Publ. No. 231, July 1995.

About the Authors

Naser El-Sheimy is a professor in the Department of Geomatics Engineering at the University of Calgary. He holds a Canada Research Chair (CRC) in mobile multisensor systems. His research interests include multisensor systems, mobile mapping systems, real-time kinematic positioning, and digital photogrammetry and their applications in mapping and geospatial information systems (GIS). Dr. El-Sheimy has envisioned, implemented, and directed the development of several commercial-grade systems for processing and georeferencing of mobile mapping data, automated 3-D mapping, and GIS applications. Dr. El-Sheimy has published over 140 papers in academic journals and conference and workshop proceedings. Dr. El-Sheimy has received many significant academic and paper awards, including the ISPRS Best Young Author Prize, the IEEE VNIS Best Paper Prize, The U.S. ION Best Student Paper award, The U.S. ION Best Paper award, the IEEE PLANS Best Paper award, and the ISPRS Best Young Author Award, and Calgary Herald – Canadian Hunter Exploration Ltd. – Petro Canada Young Innovator Award. His e-mail address is elsheimy@ucalgary.ca.

Caterina Valeo is an associate professor in geomatics engineering at the University of Calgary. She received a Ph.D. in civil engineering from McMaster University in 1998. Dr. Valeo's research areas include the impacts of forest fires and climate change on water yield and forest regeneration; modeling water resources in remote regions of the Canadian boreal forest; and developing large-scale estimates of hydrological parameters using remote sensing. Dr. Valeo is a member of various organizations and represents Canada on IAHS's Hydrology 2020 Working Group, which is tasked with providing new research directions for the field of hydrology. Dr. Valeo has authored more than 50 publications in peer-reviewed journals, conference proceeding articles, and technical reports. Her e-mail address is valeo@geomatics.ucalgary.ca.

Ayman Habib received his Ph.D. in photogrammetry from The Ohio State University. Currently, he is an associate professor in the Department of Geomatics Engineering at the University of Calgary. His research interests span the fields of terrestrial and aerial mobile mapping systems, modeling the perspective geometry of nontraditional imaging scanners, automatic matching and change detection between various datasets, automatic calibration of low-cost digital cameras, incorporating analytical and free-form linear features in various photogrammetric orientation procedures, object recognition in imagery, and integrating photogrammetric data with other sensors/datasets. Dr. Habib is the recipient of several awards, including the Talbert Abrams "Grand Award" from the ASPRS (2002). Dr. Habib has authored 100 publications in peer-reviewed journals, conference proceeding articles, technical reports, and lecture notes. His e-mail address is habib@geomatics.ucalgary.ca.

Index

The Artech House Remote Sensing Library

Fawwaz T. Ulaby, Series Editor

For further information on these and other Artech House titles, including previously considered out-of-print books now available through our In-Print-Forever® (IPF®) program, contact:

Artech House
685 Canton Street
Norwood, MA 02062
Phone: 781-769-9750
Fax: 781-769-6334
e-mail: artech@artechhouse.com

Artech House
46 Gillingham Street
London SW1V 1AH UK
Phone: +44 (0)20-7596-8750
Fax: +44 (0)20-7630-0166
e-mail: artech-uk@artechhouse.com

Find us on the World Wide Web at:
www.artechhouse.com

Learning Resources
Centre